T0313339

Economic Theory

RIVER PUBLISHERS SERIES IN MULTI BUSINESS MODEL INNOVATION, TECHNOLOGIES AND SUSTAINABLE BUSINESS

Series Editors:

PETER LINDGREN

Aarhus University, Denmark

ANNABETH AAGAARD

Aarhus University, Denmark

The River Publishers Series in Multi Business Model Innovation, Technologies and Sustainable Business includes the theory and use of multi business model innovation, technologies and sustainability involving typologies, ontologies, innovation methods and tools for multi business models, and sustainable business and sustainable innovation. The series cover cross technology business modeling, cross functional business models, network based business modeling, Green Business Models, Social Business Models, Global Business Models, Multi Business Model Innovation, interdisciplinary business model innovation. Strategic Business Model Innovation, Business Model Innovation Leadership and Management, technologies and software for supporting multi business modeling, Multi business modeling and strategic multi business modeling in different physical, digital and virtual worlds and sensing business models. Furthermore the series includes sustainable business models, sustainable & social innovation, CSR & sustainability in businesses and social entrepreneurship.

Key topics of the book series include:

- Multi business models
- Network based business models
- Open and closed business models
- Multi Business Model eco systems
- Global Business Models
- Multi Business model Innovation Leadership and Management
- Multi Business Model innovation models, methods and tools
- Sensing Multi Business Models
- Sustainable business models
- Sustainability & CSR in businesses
- Sustainable & social innovation
- Social entrepreneurship and -intrapreneurship

For a list of other books in this series, visit www.riverpublishers.com

Economic Theory

Marcelo Sampaio de Alencar

Institute of Advanced Studies in Communications, Brazil

LONDON AND NEW YORK

Published 2021 by River Publishers
River Publishers
Alsbjergvej 10, 9260 Gistrup, Denmark
www.riverpublishers.com

Distributed exclusively by Routledge
4 Park Square, Milton Park, Abingdon, Oxon OX14 4RN
605 Third Avenue, New York, NY 10017, USA

Economic Theory / by Marcelo Sampaio de Alencar.

Routledge is an imprint of the Taylor & Francis Group, an informa business

ISBN 978-87-7022-405-5 (print)

While every effort is made to provide dependable information, the publisher, authors, and editors cannot be held responsible for any errors or omissions.

This book is dedicated to my family, always present in every moment of my life, and much more important than I could possibly describe or recognize in mere words.

Contents

Acknowledgements

The publication of this book is the result of the many years of the author's work at the Federal University of Campina Grande (UFCG), at the Federal University of Paraíba (UFPB), at the University for the Development of the State of Santa Catarina (UDESC) at the Federal University of Bahia, at the Senai Cimatec University Center, Salvador, and at the Institute of Advanced Studies in Communications (Iecom).

It also ensues from the privilege of cooperating with several companies, firms and institutions, through time, among which the Brazilian Telecommunications Company (Embratel), the Technical-Scientific Association Ernesto Luiz de Oliveira Junior (Atecel), the Brazilian Company of Mail and Telegraphs (Correios do Brasil), the Telecommunications of Rio Grande do Norte (Telern), the Hydroelectric Company of the São Franciso (Chesf) and the Telecommunications of Paraíba (Telpa).

This cooperation also involved the Telecommunications of Pernambuco (Telpe), the Tele Nordeste Celular (TIM), The National Agency of Telecommunications (Anatel), Siemens of Brazil S.A., Oi, Brazilian Telecommunications S.A. (Telebrás), Alpargatas S/A, and Licks Attorneys. The experience this text mirrors to the readers is the fruit of this long-date cooperation.

I thank the exquisite translation of some of the chapters, done by Thiago Tavares de Alencar, the complete revision of the text, done by Junko Nakajima, and the elaboration of the cover, a creation of Matheus Fiqueiredo.

I also thank the administrator and consultant Romeu Soares de Alencar, Managing Partner at RS2 Consulting, for reading and commenting on the originals, and for the excellent and undeserved introduction prepared for the book.

The book was substantially improved with the critical revision of Professor Bertrand Sampaio de Alencar. Any mistakes, found perchance in the text, are exclusive responsibility of the author.

The understanding and affection of Silvana, Thiago, Raissa, Raphael, Janaina, Marcella, Vicente, and Cora, who forgave the long periods of absence on account of the academic work, allowed the author to develop this book, based on articles published in journals, magazines, and conferences, in addition to articles written by the author for the column Difusão, published by the Jornal do Commercio, Recife, Brazil.

Marcelo Sampaio de Alencar

Presentation

It was with great satisfaction and honor that I received the invitation from Professor Marcelo Sampaio de Alencar to write the introduction to his new book.

The responsibility, at first, weighed over my shoulders, not only because Marcelo is an icon in the Alencar family, composed of other brilliant minds which flourished through the centuries in Brazil, but especially for knowing his grandiose professional trajectory.

I tried to summarize his amazing curriculum, but I noticed this was the hardest part, even if I filled pages and pages, there would always be notable accomplishments and important qualities left out.

In this arduous task, I present him as an Electrical Engineer, graduated from the Federal University of Pernambuco, Master from the Federal University of Paraíba, Ph.D. from the University of Waterloo, Canada, Professor at the Federal University of Campina Grande, Visiting Professor at the University of Toronto, Canada, at the Federal University of Bahia and at the Senai Cimatec, in addition to a consultant to many national and international companies.

He is also the founder and president of the Iecom, founder and vice-president of the Paraíba Academy of Sciences, a researcher with hundreds of scientific and engineering works published, columnist for important newspapers, speaker, lecturer, and writer with 28 published books that embrace several themes, from the science of music, to probability and stochastic processes.

In addition to being a scientist, Marcelo is a poet, musician, carnivalesque, an arduous defender of his beliefs, and a militant in the search of a just and egalitarian society. I would highlight one of his qualities: Intense!

Moreover, because of this characteristic, while talking about socialist principles, defending his convictions, that the present book appeared.

He is a thinker! Now, his thought turns to the explanation of Economics, a theme that greatly interests me, for the beauty of the subject, for all its implications in our lives and daytime activities, as well as for the proximity with the activities developed in my profession.

I started to read the outline of the book thinking that I would have to read each paragraph twice or three times to be able to comprehend, such as the complexity of the author's ideas. The pleasant surprise came from the first chapter, with an analysis of the economic concepts compared to the mysteries of the universe; this great engineer demonstrated his capacity to speak about Economics in a simple manner, with a good dose of humor, and in my opinion, in a poetic way.

I found particularly interesting the correlation of the formation and behavior of the universe with the economy, and in the following chapters, I learned a great deal about the economic history of Brazil and the World.

Being a staunch capitalist, I had to open my mind to understand the profound and beautiful vision about economics, divergent from some concepts – or prejudices – I had. The book made me reflect on some paradigms with respect to the economic lines of research and it freed me from several tethers.

Romeu Soares de Alencar
Managing Partner of RS2 Consulting

Preface

"While in the economic order there is not an exact balance of forces, production, salaries, labor, benefits, taxes, there will be a financial aristocracy, that grows, shines, fattens, bloats, and at the same time a democracy of producers that slims down, withers away and dissipates in the proletariat."
Eça de Queirós

The word economy combines the Greek terms home and law, or rule; in other words, it initially meant the wise and legitimate administration of the house, for the common good of the entire family. The meaning of the term was, since then, extended to reach the administration of the State. This was the way Jean-Jacques Rousseau (1712–1778), created the concept of economy in 1755 when he presented the work "Discourse about Political Economics" (Rousseau, 1973).

Philosopher, the political theorist, writer, and composer, born in Geneva, considered one of the main philosophers of Enlightenment and a precursor of romanticism, Rousseau defended the democratic-republican ideals inspired by Sparta and centered on the idea of liberty as active participation in politics and legislation (Balckburn, 1997).

In his work on Political Economics, he analyzes the function of the State in a society with a metaphor of the human body, formed by the head and members. From this model, he shows how all political societies are composed of interests and rules, subordinated to the will of the majority.

Eventually, the term Economics started to designate the science that consists of the analysis of the production, distribution, and consumption of goods and services. In addition, it is also considered a social science, and Political Economics studies the economic activity of human beings.

That is how David Ricardo (1772–1823) would classify it, the British economist and politician and one of the most influential classic economists, considered the successor to Adam Smith (1723–1790), and one of the founders of Political Economics, for profound study of economic themes, which lead him to develop a theoretical perception of the problems of the time (Ricardo, 1996).

Organization of the Text

This book has the objective of presenting the evolution of economics, from a historical, political, and scientific perspective. The Universe of the Economy is discussed in Chapter 1, in which several economic concepts are presented in a comparative analysis with the formation of the Universe.

Chapter 2 discusses the economic formation, from the feudal system, to the formation of the State, until the present. The concept of capitalism is presented in Chapter 3, which verses on the creation of wealth, formation of companies, corruption, tax evasion, and tax avoidance.

The basic concepts of Economics are discussed in Chapter 4, including Macroeconomics, Microeconomics, Econometrics, the activities, and conditions of economic nature, as well as its division into sectors.

Chapter 5 presents the fundamental mathematical concepts for understanding economics. Microeconomics is discussed in further detail in Chapter 6, which includes themes such as costs associated with management, company capital, society formation, operations in the stock market, financial institutions and competition.

Chapter 7 presents the details of the theme of Macroeconomics, including the discussion on currency, public accounts, taxation, capital remuneration, economic policies and international transactions.

Chapter 8 discusses the relation between taxes and wealth distribution, and utilizes mobility economy as a study case.

An analysis of the economic results of privatizations, using Great Britain and Brazil as examples, is presented in Chapter 9. The chapter explains privatizations, indicating the arising problems, the consequent lack of investments, and the revenue losses incurred by the State.

The subject of Mathematical Economics is presented in Chapter 10, with an initial discussion on Econometrics, economic modeling, stochastic analysis, models for production, for the price of stocks, for the stock of companies in an Economy and for productive economic activity.

The book contains three appendixes. Appendix A presents the Theory of Probabilities, which is essential to the understanding of modern Economics.

A complete Glossary on Economics, with explanations for the most important terms, acronyms, concepts, and expressions in the area, complete the book and should be consulted by the reader, every time one wants to quickly understand a concept.

The book also contains the Biography of the Author and a Table of Contents, to facilitate the search for information in the text. Numerous bibliographic references are presented, for those that intend to specialize themselves in the area.

The text can be utilized, as a function of the course syllabus, for a term or semester long undergraduate course about History of Economics, Basic Economics, and Economics, Business, or Engineering courses.

List of Figures

1

The Universe of Economics

"Science is the great antidote to the poison of enthusiasm and superstition."
Adam Smith

1.1 Introduction

The workings of the world of Economics have a lot to do with the evolution of the physical Universe, and can be modeled in an analogous manner, if poetic liberty is allowed.

1.2 The World of Economics

Metaphysical considerations apart, the most current theory of the organization of the Universe presupposes a stage of uniform distribution of matter, composed mainly of hydrogen (Alencar, 2015c) and (Alencar, 2015b).

Economics probably originated from the activity of the human gatherers that wandered between the regions in search of food and shelter. The origin and the exact date of its beginnings cannot be precisely known, but in some singular point, or various distinct places, the production and the exchange of goods began (Alencar, 2012b).

The shocks and fluctuations between the atoms of Hydrogen may have created regions of greater mass density, with the capacity of attracting the matter, right after the start of the Universe, which resulted in the creation of matter, in a singular expansion (Rumjanek, 2009).

The establishment of an agropastoral society, with the fixation of humans to the land, changed radically the economic model. From the beginning, with a uniform distribution, wealth started displacing to the more productive, or higher demand, nuclei.

Figure 1.1   Economics looks like the Universe.

The agglutination of matter in specific points of the Universe started the formation of stars, as illustrated by Figure 1.1. The huge pressures exerted over the Hydrogen nuclei started a fusion process, which resulted in Helium nuclei, with the consequent liberation of energy. Thus began the life cycle of the stars.

Around the productive agricultural nuclei, villages of workers were established, which transformed into cities, with artisans and merchants, which diversified the economy. About this initial phase, it is worth it to remember an excerpt from the opening of the second part of Jean-Jacques Rousseau's "Discourse on the Origin and Basis of Inequality Among Men" (Rousseau, 1973):

> The first to fence a land had the audacity to say this is mine and found simple enough people to believe in him, was the true founder of civil society. How many crimes, wars, murders, how many miseries and horrors would it have saved humankind if someone had knocked down the fence and yelled to everyone: "Do not listen to this impostor! All will be lost if you forget that the fruits belong to everyone and the land to no one."

The star cycle produces, by fusion, several chemical elements, which can be expelled in spectacular explosions, known as supernovas. This process spreads heavier matter throughout the Universe.

Figure 1.2   Galaxies were formed from the agglutination of matter in the Universe.

The construction of cities allowed the coexistence of artisans, merchants, and the formation of a consumer class, for diversified products. The export of production surplus began.

The matter expelled by supernovas eventually agglutinated, to form planets and other stars, which started to revolve around the stars. The stars, in turn, started to orbit each other, forming galaxies, almost always with a supermassive nucleus, as illustrated in Figure 1.2.

The growth of cities greatly diversified the economy, with the supply of new products and the creation of specialized jobs to absorb the workers, especially those from rural areas.

Companies were created and dislocated economic power from the feuds to the burgs or cities, fostering the bourgeoisie, at the expense of feudal power. The market then appeared (Alencar, 2012b)

1.3 The Economic Universe

The concentration of mass in supermassive objects is equivalent to the formation of big companies, which swallow the small ones until they form gigantic conglomerates. These conglomerates, usually function for periods that depend on the stability they acquire (Alencar, 2012d).

Figure 1.3   Supermassive stars exploded as supernovas.

In the case of stars, a smaller mass makes them remain with a small hot nucleus, which lasts for a long time, up to billions of years, like the Sun, and ending its life as a white dwarf.

Stars with larger mass, because of the pressure exerted by the outer layers, tend to collapse faster, exploding into supernovas, in hundreds of millions of years, as illustrated in Figure 1.3.

Even more massive stars can suffer a gravitational collapse to form neutron stars, or at the limit, end up as black holes, entities with absurd mass density in infinitesimal regions of space, also known as singularities or mass impulses. A black hole can absorb anything, due to its intense gravitational pull, which nothing can escape (Gal-Or, 1981).

These singularities absorb everything around them, including light, as long as it is inside its radius of action, known as Schwarzschild radius, named due to the work of Karl Schwarzschild (1873–1916), German astronomer and physicist and one of the founders of modern astrophysics.

The companies, in this analogous model, work as the stars, with the consumers orbiting around them, like planets and comets. Big companies are similar to big stars, which are born, grow, and eventually degenerate and die.

Stars with an adequate mass density can last billions of years, similar to well managed companies, which have essential products, can last for tens or hundreds of years.

Supermassive companies can consume their elements more rapidly, and disappear in a shorter time, taking with them part of the consumer market. Several great mass companies, which seemed to be robust and eternal, quickly disappeared, such as: Kodak, Blockbuster, and Xerox.

Big conglomerates are like galaxies, which usually orbit black holes, where they might end up in. They can, in their lifetime, absorb other conglomerates, as if they are gigantic galaxies, to produce immense holdings, which can also go into the hole. The attraction forces that maintain the structure become more intense, causing its collapse. The internal forces of these entities are so intense, that not even information can escape to the market (Alencar, 2012d)

1.4 The Density of Wealth

As the cooling of the Universe lead to the formation of the stars, or vice versa, the heating associated with the end of the last ice age allowed the appearance of agriculture, with the lands that outcropped and were uncovered by the ice (Alencar, 2013a).

The economy went from the agricultural phase to the industrial phase, with the consequent accumulation of wealth. Income, that previously was associated with the leasing of lands for production, increased, permitting the accumulation of capital, the origin of capitalism.

The mass density, or in other words, mass per unit of volume, is what really matters to know if a certain part of the Universe can expand forever, stabilize, or collapse. This is related to the cosmological constant, introduced and that was later rejected, by Albert Einstein (1879–1955).

The density of resources in an economy serves a similar role. It indicates if the market has a vitality to grow, or if it will exhaust in a short period. The income per capita is the income density per unit of inhabitant. It represents the measure of expansion or contraction possibility, of the Economy of a country, for instance, by the potential to generate consumption, with the stimulus to form companies.

The amount of companies that an economy supports, on the other side, can be estimated from the Gross Domestic Product (GDP) of a country, which is the sum, in monetary values, of all the goods and final services produced in a certain region, in the period of a year.

Discounting the value lost in depreciation during the lifespan of a permanent asset, which can reach 10% of the GDP, results in the Net National Product (NNP), the measure that relates the amount of wealth in a certain nation, state or city, with its productive capacity.

Different from the uniform mass distribution, a black hole is a result of an elevated concentration of mass, like the collapse of a great conglomerate, which has the potential of bankrupting a fair share of the world economy, causing recession (Berthoud, 2000).

Currently, in the macroeconomic analysis there is a tendency to use Purchasing Power Parity (PPP), in which the goods that must be sold at the same price in any country are considered. This implies that the nominal exchange fee reflects differences in price levels. This parity adjusts the GDP and the per capita income of the poorer countries, considering that the services and goods in those countries have lower costs.

The general level of economic activity in a capitalist system, according to John Maynard Keynes (1883–1946), economist and employee of the British treasury, is determined by the disposition of the entrepreneurs in doing capital investments (Heilbroner and Thurow, 1982). Keynes, appearing in the Figure 1.4, was one the greatest economists of the 20th century.

Keynes founded modern macroeconomics, perceived the unstable nature of capitalism and its incapacity to promote the well-being of the society. His basic ideas include the opposition to the liberal and neoliberal ideals, a certain level of protectionism and economic balance, and government capital investment. The reduction of interest rates, a balance between demand and production, state intervention in the economy, the guarantee of full employment, and social benefits (Roncayolo, 1973).

Ernst Friedrich Schumacher (1911–1977), a German economist and statistician, was a follower of Keynes' theory, and developed an economic theory based on fighting waste and preserving ecology. A curious fact about his family, relating to the marriage of his sister, Elizabeth Schumacher, with the famous physicist Werner Heisenberg (1901–1976), one of the pioneers of Quantum Mechanics, and better known for Heisenberg's Principle of Uncertainty, which he actually called, the Principle of Imprecision, published in 1927 (Schumacher, 1990).

Schumacher said that from his experience, the principles of socialism applied were better applied to non-economic values, but he was a scathing critic of entrepreneurs, which, according to him, became primitive, especially by the process of reducing thousands of aspects of the economy in the search for money. They fit into this simplified version of the world and are

Figure 1.4   John Maynard Keynes was one the greatest economists of the 20th century and created modern macroeconomics.

satisfied with it, but get confused and helpless in addition to feeling exposed to unstable forces and uncertain dangers, when the real world occasionally show one of its different faces.

After analyzing the pros and cons of the capitalist and socialist systems, Schumacher reached an intermediary system, to characterize the private property of enterprise, based on the work of Richard Henry Tawney (1880–1962), a British economic historian, social critic, and Christian socialist.

1. For a small-scale enterprise, private property is natural, fruitful, and just.
2. In the case of a medium scale company, private property is, for the most part, functionally unnecessary. The idea of property becomes tense, unfruitful, and unjust. The disposal of privileges for a larger group of workers is advised.
3. For large-scale companies, private property is a fiction with the intent of allowing owners with no function to be a parasite of other's work. This

is not only unfair but an irrational element that distorts the relations in the company.

1.5 Singularities in Communications

In Communications, as in Physics, there is an entity known for its unorthodox or unconventional properties – the impulse, or singularity. The impulse only exists for an infinitesimal amount of time. The time between a traffic light turning green and the honk of the horn from the car behind you, for instance, but has infinite amplitude, despite having a strictly finite area.

The impulse function models several electrical and physical phenomena, such as discharges from capacitors, inductors, lightning, sampled signals in communications, black holes. Even the great poet Vinícius de Moraes gave a poetic definition for the impulse, in the poem "Happiness Sonnet" (da Cruz e Mello Moraes, 1946):

> "Let it not be eternal, since it is a spark, but let it be infinite while it lasts".

Black holes are well-done examples that nature offers for the impulse. They are the result of the collapse of gigantic stars, which after the supernova phase implode to form superdense entities, despite the finite mass. The black holes absorb everything around them. Anything that reaches their event horizon can never escape, as illustrated in Figure 1.5.

Figure 1.5   Supermassive companies can consume the economy forming black holes.

Many communications companies had their star phase, some even became supernovas, others white dwarves. The Radio Corporation of America, best known as RCA, reigned in the United States as one of the biggest companies of all time.

Piloted by David Sarnoff (1891–1971) for over fifty years, RCA influenced American life, movies, music, television, radio, as well as the market. Even the Federal Communications Commission (FCC) needed to have his endorsement. It withered away after the exit of its commander until it became a white dwarf and had its brand acquired by the Korean company Samsung.

In Brazil, Telebrás was once a giant monopolist and had its supernova phase during the military regime, obfuscating the glow of Embratel, another very big company. It generated, after its explosion by the government, tens of other companies all around the Country, until regressing to the white dwarf stage, remaining currently undead in the limbo of governmental bureaucracy.

WorldCom, a gargantuan conglomerate of over 60 companies, had its stellar explosion almost two decades ago. It lived the euphoria of the unlimited markets, burning in its colossal furnace whatever was in its radius, amalgamating with however many companies were necessary to become the largest in communications in the world.

However, unlike Telebrás, WorldCom had enough mass to become a singularity, a company that reaches the impulsive state and can drag to the black hole of its bankruptcy a fair share of the companies that are under its control. Embratel had to fight to stay away from its event horizon.

1.6 Singular Wealth

The gravitational singularity is a concept in Cosmology that involves the infinite curvature of space-time, or in other words, defines a point in which the mass density of a body is so elevated, that the curvature in space-time caused by it is infinite. This is a function of the extremely intense gravitational field produced in the surroundings of the singular point (Alencar, 2018a).

In Mathematics, the singularity is a point in which the function assumes infinite values, or has an indefinite behavior. This singularity is also known as the impulse function, initially described by the British mathematician Paul Adrien Dirac (1902–1984), for calculating probabilities.

The singularity associated to gravity is better known as a black hole, a term created, in 1968, by the North-American theoretical physicist John Archibald Wheeler (1911–2008), to describe in a dramatic manner the most perceptible characteristic of the phenomenon, in other words, light cannot reflect or escape from inside it. Therefore, the black hole cannot be seen.

The idea of a body with great mass, from which nothing can escape, not even light, was initially formulated by the British geologist John Michell (1724–1793), in a letter written to Henry Cavendish (1731–1810), of the Royal Society, in 1783.

In 1796, the mathematician Pierre-Simon Laplace (1749–1827) mentioned the idea in his book "Exposition du Système du Monde", in which he used the terminology "black star," a sort of conceptual ancestor of the black hole.

The modern concept of a black hole is due to the work of Albert Einstein, since his General Theory of Relativity predicted its formation. However, the first scientist to demonstrate that these structures can form was the German physicist and astronomer Karl Schwarzschild (1873–1916), in 1915, in a letter to Einstein.

The black holes are formed from the collapse of supermassive stars, with masses vastly superior to the Sun, and continue to suck all the bodies around them, being able to consume an entire galaxy after a while.

At the center of the Milky Way, the galaxy which contains the Sun in one of its arms, there is a black hole whose gravitational force compels all the bodies to orbit, into a kind of macabre dance, until they are finally swallowed in a colossal feast that will consume the entire galaxy in a few billion years.

The known universe, in its initial period of life, had a fairly uniform mass distribution, and it seemed like it would continue that way indefinitely.

However, little localized fluctuations increased the density of the existing Hydrogen, causing the ignition of the stars, which started to produce, by fusion, other heavier elements.

The fusion continued until the synthesis of Iron, which, instead of producing, absorbed energy. This breaks the delicate balance that maintains the stars functioning. The production of Iron caused the collapse of the massive stars, creating supernovas, gigantic explosions which lead to the formation of other natural elements of the Periodic Table, which originated the planets and other celestial bodies.

The gases, especially Hydrogen and Helium, together with cosmic dust, formed by other chemical elements generated in the supernovas, known as the star nurseries.

The stars, with eventual planets, asteroids, and comets around them, compose the galaxies, which can agglutinate, to form new, larger, and more complex galaxies. The Milky Way with 200 to 400 billion stars is, possibly, the result of the fusion with the galaxy Gaia-Enceladus, 10 billion years ago.

The world wealth distribution works in a similar fashion to the mass distribution in the Universe. In the beginning, the economy was simple and relatively homogeneous, with hunters, miners, and farmers exchanging goods, thousands of years ago.

This propitiated competition and free initiative, since the economic power, the so-called market power, was distributed. However, due to small fluctuation in the trade or power relations, the wealth started to concentrate in certain groups, some segments of the economy, or in few hands.

Since then, the accumulation of wealth has been the keynote, leading to fusions and acquisitions, which fed the current economic model. This accumulation is harmful to the economy, because it causes the search for products with high prices and little utility, causes inflation of demand, removes resources from the productive economy, drains available money.

As a result, according to data collected by the French economist Thomas Piketty (1971-), the Money applied only in obtaining income, without a direct link to production, is already over six times the production of all the countries put together (Piketty, 2014).

This can seem strange or absurd, but there is no mistake in the previous statement. The rentiers today have an income equivalent to 600% of the GDP of the World, in other words, this parcel of capitalists could buy the entire production of goods and services in the World for six years in a row, leaving the rest of the population with literally nothing.

Since the world did not improve this capitalist accumulation, the theory developed by the Nobel Prize recipient Simon Kuznets (1901–1985), Byelorussian, naturalized American, an economist, seems today only a fairy tale. According to Kuznet's theory, elaborated in 1955, the income inequality should diminish in the more advanced stages of capitalism, independently of the economic policies adopted or the differences between the countries, in opposition to the predictions of Karl Heinrich Marx (1818–1883), done a century earlier (Piketty, 2014).

Pleonexy was the term that Plato (428a.c.-348a.c.), founder of the Academy, and his disciple Aristotle, creator of the Lyceum, used to designate the basis of injustice in the society. The term means having more, acquiring more, without the motivation of necessity.

These philosophers saw social inequality as the core of all injustice and, for this reason, coined the term to designate the lust for gaining, greed, which, according to them, lead society to problems such as addiction and injustice.

Adam Smith (1723–1790), also a Philosophy professor, had a similar vision in relation to economic inequality.

According to Aristotle, in distributive justice, in which the sharing of goods between people is processed according to the merit of each one, being unjust consists in wanting to possess more, in search of an undue advantage, taking from another person what that person deserved.

The current pleonexy leads the rentiers to hold, solely in the United States, US$ 130 trillion, six times the largest GDP in the World, of a little over US$ 21,5 trillion. They are colossal black holes in the economy, which suck everything and everyone, in an unrestrained accumulation, which consumes everything, destroys lives, and produces little.

1.7 In the End it will be Chaos

The chaos of the Greek or Christian mythologies has served the creation of many physics theories to explain the workings of the World. In the beginning, there was the movement due to the clash of two rigid bodies, of spherical format. This movement is predictable by the laws of Physics, and the formulas from the deterministic model appear in high school books (Alencar, 2014g).

However, when three or more bodies clash, the result is susceptible to small variations in any of the variables involved, such as speed and the amount of movement. The result can be chaotic.

For this reason, Albert Einstein (1879–1955) developed the mathematical theory of Brownian movement, which resolves the problem of clashes between bodies, as long as their numbers are considered very high, and that the variables involved are random (Einstein, 1905).

In other words, randomness was introduced by Einstein in Physics to eliminate the chaos, which seems surreal. But makes a lot of sense, when considering that the Theory of Probability allows results to be obtained on an average, which is perfectly reasonable and acceptable for systems that have an absurdly elevated number of components.

Chaos theory is also used to explain phenomena such as lightning, atmospheric discharges which hit Brazil more than any other country, in which the accumulation of charged particles, probably due to atmospheric movements, which dislocate the electrical charges, amid the Earth's magnetic field.

Chaos also models electrical blackouts, since a small variation in voltage or current at some point of the power system can evolve into an overload, and result in the activation of the protection equipment, which cuts off the lines.

Chaos is also explored in communications. A chaotic modulator uses a carrier with a chaotic phase or frequency. The combination of two sines whose frequencies are related to irrational numbers generates a carrier with a chaotic characteristic. The use of this type of carrier allows the protection of the transmitted signal against undesired wiretaps.

Lastly, but not exclusively, chaos is used to describe an economy that does not obey the usual precepts, such as the law of supply and demand. The economy is said to be chaotic, as is the characteristic used to describe Brazil since time immemorial.

The Universe can be unstable, depending on the gravitational constant proposed by Albert Einstein. It can expand until a point in which local fluctuations lead to elevated concentrations of mass, which attract other masses, uncontrollably, until the formation of a colossal black hole, with the mass of the Universe itself. This would be what was agreed upon to be called the big crunch – and it represents the end of the current Universe.

In a similar fashion, the liberal model which, in the absence of State control, allows excessive income concentration, in which the most profitable companies buy other less profitable ones, can lead, in the limit, to an economic singularity, in other words, all the resources appropriated by a single company, or investor, and no income distribution in the population.

This theoretical model would lead to the general impoverishment of the population, in detriment to the enrichment of a single investor – and the current economy would end.

1.8 When Light Bowed to Ceará

In the analogy utilized in this chapter, it is important to remember that gravity is considered the weakest of the fundamental forces, usually called interactions in Physics, but is the one with the longest reach. The concept of reach refers to the decay of the interaction with distance, in other words the force decreases with the square of the distance between interacting objects.

However, distinctly from other interactions, gravity acts in a universal manner in all matter, such as planets and stars, but also in the energy present in space, as the deflection of light by objects with great mass demonstrates. Einstein predicted and verified this deflection during a solar eclipse on May 29, 1919, accompanied by a team of British astronomers in the city of Sobral, State of Ceará, Brazil.

Figure 1.6   Delegation of British scientists, together with the population of the city of Sobral, waiting for the eclipse. Image adapted from Wikimedia Commons, a collection of free content from the Wikimedia Foundation.

Figure 1.7   The British scientists use a 13-inch telescope to observe the eclipse. Image adapted from Wikimedia Commons, a collection of free content from the Wikimedia Foundation.

Figure 1.6 shows the delegation of British scientists, together with the population of the city of Sobral, ready to observe the astronomic phenomenon that would change the course of Physics and would put Einstein definitively on the list of the greatest scientists in the world (Pivetta and de Oliveira Andrade, 2019).

The delegation of British scientists used a 13-inch telescope to observe the eclipse, as shown in Figure 1.7. The astronomer Henrique Charles Morize (1860–1930), director of The Brazilian National Observatory was the responsible for sending the information about the favorable geographic conditions for the observation of the phenomenon to the Royal Astronomic Society of London (Alencar, 2008c).

Due to its long reach, and because of the sole dependence of the mass of objects, the most part of the interactions between objects separated by distances larger than the dimensions of a planet is caused by gravity.

Gravity is responsible for large-scale phenomena, such as galaxies, black holes and the expansion of the universe, in addition to the prosaic astronomical phenomena such as the orbit of the planets, or daily experiences with falling objects or stock markets.

2

History of the Economic Formation

"To be radical is to grasp things by the root."
Karl Heinrich Marx

2.1 Introduction

Without the state, there are no rights, just like in the feudal regimes, in which people only had the duty of maintaining their lords with their work, and did not even have the right to live, a prerogative of the feudal lords (Alencar, 2016a).

2.2 The Ancient Feudal System and the Formation of the State

The democratic state was created to guarantee the rights of the citizens against the absolutism of the feudal lords, the monarchs, the pharaohs, the religious leaders, and the emperors. The landowners were, at the same time, the governors, legislators, judges, and religious leaders.

It is not strange that the first constitutions emphasized so much the right to life and the individual guarantees in their texts. Because without the democratic State, that right would not exist.

The *magna cartas* also insisted on the separation of powers, one of the individual rights guarantees, like the separation from the church, to promote the secular and democratic state.

Feudalism was predominant during the entire Middle Ages, a precursor of the bourgeoisie, birthplace of capitalism, and was characterized by vassalage and authority relations and land ownership of the feudal lord. The villages became the center of the socio-economic structure, with a productive system directed to the provision of the individual necessities of the feuds.

The feuds were a territorial unit of the feudal economy, with the characteristic of economic self-sufficiency, agro pastoral production, and the absence of commerce.

Nobility, composed of the feudal lords, had the main function of warring, in addition to exercising their political power over the rest of the classes. The monarch gave out the land, and the nobles gave military support.

They were typically divided into three areas. The domain, where the fortified castle was erected, was an exclusive right of the feudal lord, but was worked by servants.

The holding, designated to the servants, was divided into tracts, of which usually half of the production was allocated to the feudal lord. The common lands, the communes, encompassed the woods and pastures that could be used by the lord or the servants.

The clergy, despite being important in the feudal society, was not a separate class, because the components of the clergy either were lords, the so-called high clergy, or they were servants, integrating the low clergy.

2.3 The End of the Feudal System and the Start of Democracy

The end of the feudal system, that originated urban economies, the bourgeoisie, coincides with the fall of the Oriental Roman Empire, the defeat of Constantinople, in the 15th century, due to several economic, social, political and religious reasons (Alencar, 2015b).

Among the reasons, the principal was starvation, caused by the stagnation of agricultural techniques allied to the excessive population growth, the plague, which ravaged Europe decimating a third of the population, the depletion of mineral reserves, which reduced the coin production, affecting the banking operations and commerce, in addition to the rise of the bourgeoisie.

Servants constituted the majority of the peasant population. They were tied to the land, exploited, forced to serve the nobility, and pay several tributes in exchange for the use of the land and military protection. It was a miserable life, despite the servants not being slaves.

Due to the expropriating characteristic of the feudal system, the servant did not have an incentive to increase production with technological innovations, since the lord would take any surplus produced.

The servants had many obligations to the feud. Among the tributes there was the *corvée*, the compulsory labor in the lord's lands, a few days a week, the *taille*, in which part of the servant's production was given to the noble, usually a third of the production, called the banality.

Curiously, there was a tribute charged for the use of instruments or goods belonging to the feud, like the mill, oven, barn, bridges, capitation, a tax paid by each family member, the tithe, in which 10% of the servant's production was paid to the Church. The census, a tribute that the villains, the free people of the village, should pay, in money, to nobility, the justice tax, paid by servants and villains to be judged in the noble's court.

There was also the formariage, that every servant was forced to pay to help in the marriage of the feudal lord or his family member. Marriages between servants had to be accepted by the noble, which in some regions, could request to enjoy the female servant's wedding night in place of the official husband.

As if it were not enough, there was still the mortmain, the payment of a tax for servant family to stay in the feud, in the case of the death of the father or the family. Lastly, the servant had the obligation of housing the feudal lord, a duty known as lodging.

The complaints about the current amount of taxes seem childish when the amount of taxes imposed on the servants of the feudal system are analyzed. Talking about rights in that time would summon the harshest penalties in the system.

2.4 The Development of the Post Feudal Economy

Feudalism as a political, economic, and social system, predominated in Occidental Europe from the start of the Middle Ages to the creation of the modern States, with its apogee between the 11th and 13th centuries.

In the 13th century, with the appearance of new powerful centralized monarchies, urbanization in Europe, the acceleration and diversification of the economy, among other factors, the system began its decline, which increased in the 14th century with the emergence of the bourgeoisie, and the substitution of part of the servants for salaried laborers.

Despite the cultural changes, some feudal institutions, such as the feud, lingered in Europe even after the end of the Ancient Regime, the social and political aristocratic system that was established in France.

The commercial and urban renaissance weakened the nobles in Europe and strengthened the power of the kings, which had little authority in the medieval period. In some countries, the sovereigns had the support of the bourgeoisie, that was interested in political centralization because the standardization of weights, measurements, and currency and the unification of justice and tribute favored commercial development.

The nobility, after some conflicts, did not have an alternative other than accepting the royal dominance, and was, in part, coopted by the formation of courts, assembled by nobles supported by the State. Kings obtained political, economic and military control of countries, which lead to absolutism.

During the 17th century, the Ancient Regime started to decline due to Enlightenment, the chain of thought that defended the liberalist ideals, such as a the institution of a manager subordinated to a constitution. Enlightenment also preached the end of interventionism, political and economic, the universal vote and democracy.

Furthermore, the Industrial Revolution and the bourgeoisie assumed an elevated social position, and started requiring a representative of their interests in the government, which weakened the absolutist system. Slowly, the monarchy was losing power, particularly in England, due to the political and social advancements produced by the first industrial revolution.

The modern economic theory emerged with Enlightenment. Considered the father of modern economics and the most important theorist of economic liberalism, Adam Smith (1723–1790) was a British professor, philosopher and economist, born in Scotland. Figure 2.1 shows a portrait of Smith, probably painted after his death.

Smith, that studied the economic relations in Great Britain in the three centuries preceding the publication of his most famous book, An Inquiry Into the Nature and Causes of the Wealth of Nations had a certain aversion to mercantilism, and his most famous quote, featured in the book, "So, the merchant or salesperson, moved only by his own selfishness, is led by an invisible hand, to promote something that never was part of his interest: the well-being of society." was, actually, criticism of the greed of merchants (Smith, 1986).

The expression of the invisible hand, which is analyzed to exhaustion by the liberals and capitalists, first appears in the essay that Smith wrote on the history of astronomy. Later, it reappears in the Theory Of Moral Sentiments. Despite their cupidity and insatiable rapacity, the rich are in fact incapable of consuming a lot more than other people and, thus, are taken by the invisible hand to do "approximately the same living needs distribution that would have been done if the Earth had been divided into equal portions among all its inhabitants" (Balckburn, 1997).

Smith was tolerant to state intervention in banks, to combat poverty, and to promote equity, if the State regulations favored the laborer. He was also in

Figure 2.1   Adam Smith was the greatest Scottish philosopher and economist. Image adapted from Wikimedia Commons, a collection of free content from the Wikimedia Foundation.

favor of national currency, regulated by the State and emitted as a function of the surplus value and not depending on the debt. He affirmed that the main problems in economic relations were the lack of probity and punctuality, which lead to a crisis of confidence and to this end defended the regulation of the financial market.

David Ricardo (1772–1823), British economist and politician, exerted great influence over both neoclassical economists and socialist economists, and his work was important to the development of economic science. His works include studies on the value-labor theory, the theory of distribution, which discusses the relations between profit and salaries, international commerce, in addition to monetary themes (Ricardo, 1996).

Ricardo, shown in Figure 2.2, treated the distribution of the product generated by society's labor. In other words, the analysis of the joint application of labor, machinery, and capital in the productive process to generate a product.

This product would be divided among the classes in the existing society of the time, the owners, which earned income from land, the wage-earning laborers, that would receive payment, and the capitalists, that would receive profits from the capital.

Figure 2.2   David Ricardo was one the greatest British economists and politicians. Image adapted from Wikimedia Commons, a collection of free content from the Wikimedia Foundation.

The role of economic Science would be, in his view, determining the natural laws to orient that distribution, as a way of analyzing the perspectives of the economic situation, without losing the concern with long-term growth.

Ricardo's theory of comparative advantages forms the basis for the international commerce theory. He demonstrated that nations could benefit mutually from free trade, even if a nation is less efficient in producing goods as their commercial partners.

Ricardo defended the thesis that the amount of money in a certain country, as well as the monetary value of this money, would not be determinant to the wealth of the country. According to him, a nation is rich because of the abundance of goods that contribute to the commodity and well-being of its inhabitants.

2.5 The Current Feudal System

A few decades ago, the elimination of citizens' rights, like in the case of laborers, for instance, with the reduction of the state, to the so-called minimum state, is the mantra of a political chain called neoliberalism (Alencar, 2015a).

This right-wing branch caused the sale of state-owned companies in many countries in the last decades, a perverse process of privatization of public assets that was the genesis of the corruption process that happens today.

In Brazil, the resources obtained indirectly with privatizations guaranteed resources to approve the amendment of reelection, and the election of the president who proposed the privatization of the State companies.

But that is not all. Capitalists, heirs of the feudal lords, intend to, with the minimization of the state, possibilitate the unlimited exploitation of laborers, without apprehension of punishment by fragilized states.

The current feuds are not restricted to the lands of the countries. There are private feuds in the telecommunications area, the automotive sector, in the energy sector, media, mining, educational, financial urban waste, and health sectors.

Those new feuds are composed of oligopolies, over which the State has little or no control. The telecommunications companies, for instance, stipulate and charge high prices, offer poor services, and despise the rules established by the regulators, regarding the blocking probability and unavailability.

Moreover, nothing happens to those companies that do not deliver. Instead, there are studies to increase the concession periods, as was done in relation to the electric power companies, that were granted decades of public concession without being bothered, rather than the period initially established. Feuds that became, little by little, eternal.

In addition, there will soon be feuds in the petroleum and gas, waters and sewage, prison, and justice areas. Currently, there is a historic retrogression of many centuries happening; curiously, it is called progress by the neoliberals.

These days, the users are the villains, as the inhabitants of villages were called in the past, which lived under the scrutiny of the feudal lords.

The new feudal lords are the owners of huge companies, that integrate modern feuds, are not content with all the profit they have, and still cut deals and assemble gangs to rob the State.

2.6 About Tulips and Virtual Currency

The collapse of the tulip market in the Netherlands, also known as Holland, completed 380 years. It was considered the first economic bubble registered in the world and its cause was the speculation around the price of the tulip bulb, that went from a trend to investment, with its price reaching ten to fifteen the salary of a skilled artisan, at the time (Alencar, 2017b).

The tulip was introduced in Europe from Vienna, where the first seeds were planted, from Turkey, in 1554. However, their popularity was only established after the botanist Carolus Clusius (Charles de l'Écluse, 1526–1609) created a garden in the Leiden University, with a tulip plantation.

The Netherlands had become independent of Spain, and had very lucrative trade routes with the Indies. This age of affluence allowed tulips to become a luxury article, with many varieties in the market, and prices that were elevated each day by the demand of the higher classes.

The avid buyers started to sign contracts in notary offices for the advanced purchase of the bulbs, creating the so-called future market, virtual and marginal to the official market of stocks and goods, with deals being closed in taverns, with the payment of a wine tax, that corresponded to 2,5% of the negotiated value.

Because of that, the tulip became the fourth biggest export in the Netherlands, the prices skyrocketed, with bulb purchase and sale contracts changing hands up to ten times a day. The collapse started with the lack of buyers in an auction in the city of Haarlem, and spread throughout the country, bankrupting many speculators, and people in good faith.

In 2009, a character called Satoshi Nakamoto, whose real identity continues to be unknown, developed the reference for the implementation of an encrypted virtual coin, known as Bitcoin (BTC), that uses a process of data sharing among several parts, which is apparently inviolable, known as the blockchain.

Bitcoins are negotiated and stored in a decentralized network of computers, which is not under the control of any government or company. An attractive aspect for investors from countries with a history of savings, accounts, and assets sequestration; but also an obscure port for tax evaders, drug dealers, and corrupt people in general.

For those who like history, and value their assets, it is worth knowing that the virtual crypto coin Bitcoin reached the stratospheric value of US$ 63 thousand, in April 2021. That is correct, a single Bitcoin, a binary encrypted code with no real currency backing, was worth 200 to 250 salaries of a specialized technician. For the record, the alleged creator of the currency disappeared in 2010.

3

On Capitalism

"Economics is extremely useful as a form of employment for economists."
John Kenneth Galbraith

3.1 Introduction

This chapter exhibits some basic ideas regarding the history of Economics, originating with the seminal works of Adam Smith, and later developed by David Ricardo, Karl Marx, John Kenneth Galbraith, John Maynard Keynes and other philosophers and economists.

3.2 Capitalism for Beginners

Adam Smith (1723–1790) was the greatest Scottish philosopher and economist, inventor of modern Economics considered the most important theorist of economic liberalism, and author of the classic book "An Inquiry into the Nature and Causes of The Wealth of the Nations" (Smith, 1986).

In the work, published in 1776, he intended to demonstrate that the wealth of the nations resulted from the action of individuals that, moved also by their self-interest, promoted the economic growth and technological development of their country. The book contains criticism towards the United Kingdom's protectionism at the time, which Smith blamed the increase of the price of products on and a radical critique of the mercantilist theory (Bresser-Pereira, 2019).

One of his most famous quotes was "So, the merchant or salesperson, moved only by his own selfishness, is led by an invisible hand, to promote something that never was part of his interest: the well-being of society." This action would be, then, responsible for the reduction in prices and improvement of work conditions. The phrase also reveals that Smith probably did not think highly of merchants.

Adam Smith studied the economy of the United Kingdom thoroughly, since the time of Queen Elizabeth I, and concluded that closing the English ports to international commerce, as was requested by the producers of that country, was responsible for the increase in prices that followed. In other words, the competition with other countries was what maintained prices stable (Alencar, 2013b).

He also deduced that, in spite of the devaluation of the currency, caused in part by the subtraction of metal by the monarchy, or the corresponding increase in prices, the relative costs of goods remained the equivalent for three centuries. In other words, a kilogram of beans could always be exchanged for the equivalent in wheat, as long as there were no problems in production or importation.

Those that consider Smith to be a cynic philosopher may notice an essential contradiction in his biography, because of the publication of his first work, in 1759, "The Theory of Moral Feelings," which he considered to be superior to "The Wealth of Nations" (Heilbroner, 1986).

In his work, Smith examines with a critical approach, the moral thinking of his time, and concludes that conscience emerges from social relations. He aims to explain the human capacity of forming moral judgement, in spite of the natural tendency to self-interest. Smith proposed a sympathy theory in which the act of observing others makes people conscious of their selves and the morality of their behavior.

David Ricardo (1772–1823) is considered the successor to Adam Smith in the task of divulging Political Economics, a nascent science at the time. The work of Ricardo, whose family was of Portuguese origin, embraces many economic themes, which vary from monetary politics, profit theory, land income and distribution, to the theory of value and international commerce (Ricardo, 1996).

In Ricardo's vision, the landowners would appropriate a burgeoning part of the production and income. He was concerned with the evolution of land prices and its income, and was influenced by the model created by Thomas Malthus' (1766–1834), a British economist, considered the father of demography because of his theory relative to population control.

For reverend Malthus, an Anglican pastor, son of a rich landowner, the increase in population was the biggest threat at the time, considering that it would have a geometric progression, while the food production would progress arithmetically. Therefore, any increase in the living standard of the proletariat would be temporary, since it would cause an increase in population, restricting any possibility of improvement.

Because of that, Malthus imagined a global catastrophe associated to super population, which would lead the World to chaos and misery. He argued that any measures of assistance to the poor would need to be suspended and that the birthrate should be controlled strictly in this segment of the populace (Piketty, 2014).

Ricardo was more interested in the possibility of land scarcity in relation to other goods, according to the law of supply and demand; it should increase its price continually. At the limit, landowners would receive ever-growing income, in relation to the rest of the population, what would destroy the social and economic balance.

His proposed solution was to establish an increasing territorial tax, to warrant control of land possession. The price of arable land actually increased for a long period but eventually stabilized as agriculture lost importance in national wealth.

He was strongly influenced by Smith's main work and formulated one of the first versions of the Quantity Theory of Money (QTM), which forms the basis for other orthodox doctrines that combat inflation. According to the QTM, the general price level has a proportionality relation to the amount of goods and services negotiated in the economy and to the quantity of money in circulation, taking into account the payment habits of the community.

In an interesting note, during the validity of commodity currency, as civil money or mercantile money, it would materially impossible to mention inflation with the meaning of physical expansion of the currency in circulation in precious metal or in the meaning of price increase (de Carvalho, 2012).

The money would also be the form, independent of value, which possesses attributes and has functions. The State would not determine those attributes; it would only set their quantitative expression, and formalize it. There would be two attributes in money: the measure of value, the means by which the value is expressed, and the means of circulation, from where the currency is originated. The basic functions would be to act as a means of payment, operate as a means of hoarding or as a Global currency (de Souza Silva, 2018).

In one case, since the historic rarity of noble metals, used in coinage, marked the trajectory of mercantile money more because of its scarcity, then because of its abundance. In the other case, because, as time goes by, in any period, the price increase had as an immediate cause, scarcity, real or manipulated, through a monopoly of the local market.

Karl Heinrich Marx (1818–1883) was a philosopher with a bachelor's degree in Law, a historian, and an economist. Marx presented his thesis at the University of Jena, in 1941, supervised by Bruno Bauer, on *Differenz der demokritischen und epikureischen Naturphilosophie*, or "The Difference Between the Democritean and Epicurean Philosophy of Nature."

As happened with all humanist economists, Marx was also influenced by Adam Smith. However, liberal economists are accused of appropriating Smith's ideas in defense of their own cause, despite knowing of his inclination to humanism. The liberals, whose proposals greatly appeal to many entrepreneurs, preach the elimination or reduction of the government action in the economy (Balckburn, 1997). (Marx, 2004).

Marx, shown in Figure 3.1, was also influenced by Georg Wilhelm Friedrich Hegel (1770-1831), a German philosopher, recognized the social division between the classes of exploited and exploiters, noticed the manipulation to disguise this basic dissent, and imputed to the latter the task of transforming the social structure that perpetuates inequalities. The socialization of the means of production would be the way to eliminate those social differences (de Souza, 2017).

Figure 3.1   Karl Marx was the philosopher and economist that had the most influence on the World. Image adapted from Wikimedia Commons, a collection of free content from the Wikimedia Foundation.

In Marx's time, the most memorable fact was the misery of the industrial urban proletariat, due to the rural exodus caused by the increase in population and agricultural productivity. The workers, including children, worked long hours, for low salaries, and lived in tenements, despite the accelerated economic growth in the period, due to the Industrial Revolution (Piketty, 2014).

Therefore, Marx proposed an alternative to socialism, as the doctrine that recommends the organization of an egalitarian society, free of exploitation relations between social classes, and that secures the primacy of the collective over the individual interests. To this effect, a planned economy was necessary, as an alternative to the randomness of the market (Bornstein and Fusfeld, 1970).

Eventually, all the countries, socialist or capitalist, adopted economic planning, since the governments could not indulge in operating a *Laissez-faire* regime left to the whims and indecision of the market. Curiously, today the United States has one of the best and most elaborate economic evaluation systems and keeps the economy carefully planned.

However, as we know, the contempt entrepreneurs have for government intervention in the market grows when the market is effervescent, but quickly disappears when the market collapses and threatens their businesses. That's the moment when they all run to the government, cup in hand outstretched.

In some aspects, Marx's philosophy was a continuation of Hegel's dialectic, which he considered a powerful logic tool to demonstrate social laws. Marx valued scientific analysis, in a time when religion was dominant and was classified as a materialist when using dialectic to support his social progress theory, establishing higher moral values.

Marx's philosophy differed from Hegel's because he believed in the class struggle as a driving force for social change, while Hegel thought of the struggle between nations. Furthermore, Marx glimpsed power as being economic, instead of political, as Hegel thought. Neither of them believed in the power of legislation to remedy economic abuses, which were glaring in their respective times.

Marx hoped that his revolutionary movement would lead to a form of socialism, with social equality and liberty. This, in a period in which a large part of the population in the World still lived in actual slavery, or in some system of economic exploitation, in which men, women, and children had to work up to sixteen hours a day to survive (Sabine and Thorson, 1973).

Marx's social development theory, according to him, would have affinities with organic evolution, developed by Charles Darwin (1809–1882), in his

Figure 3.2   Charles Darwin wrote a seminal book on the origin of the species, and influenced other areas. Image adapted from Wikimedia Commons, a collection of free content from the Wikimedia Foundation.

magnificent book "The Origin of Species." In the book, Darwin proposed that human characteristics were hereditary. He also published "The Expression of the Emotions in Man and Animals," in which he suggests that behaviors are evolutionary adaptations (Darwin, 2009).

Darwin's studies, apart from Adam Smith's results, were also taken advantage of, in a negative manner, by economists adepts of *Laissez-faire*, known today as neoliberalism, which preaches the elimination of the State, governments and social guarantees obtained by citizens, in favor of a free market economy. Obviously, there is no free market when a few capitalists dominate it (Alencar, 2014d).

Capitalism is the economic system based on the private property of the means of production and their utilization with profit in mind. It is a regime grounded in the dissociation between workforce, or proletariat, and the means of production that are used with the fundamental objective of providing a profit. Karl Marx imagined a process of sequential revolutions that, in the end, would eliminate social classes and the exploitation of one human being by another.

In capitalism, employees execute production in Exchange for a salary that remunerates their workforce. Evidently, pure capitalism is only hypothetical, a regime in theory. Even in what is considered the mecca of capitalism, the United States, a lot of workers hold partial control of the companies, by means of shares, and an appreciable parcel of the economy is controlled by the State, with 27% of the GDP, in 1991 (Vinod Thomas, 1991).

John Kenneth Galbraith (1908–2006), an American economist from a Canadian Family, the defender of liberalism, published several Works in which he extended the idea that there could not be a free market if few companies controlled it, and criticized the assumption that an ever-growing material production was a sign of economic health (Galbraith, 1984).

Galbraith argued that price competition remained the dominant social control force in the market system composed of the majority of the companies. However, in the industrial sector, composed of a smaller set of gigantic corporations, the main function of market relations would not be to contain the power of those conglomerates, but to serve as an instrument to implant this power.

Thus, the power of those corporations would extend to politics and to commercial culture, allowing them to exert considerable influence in social attitudes and the population's judgments of value, which would be inconsistent with democracy, and a hindrance to obtaining the quality of life that the acquired affluence could provide.

In Brazil, Celso Monteiro Furtado (1920–2004), the greatest Brazilian economist and one of the most distinguished intellectuals of the country in the 20th Century analyzed the fundamentals of economic development and underdevelopment and published one of the most important economic works, "Economic Formation in Brazil" (Furtado, 1998).

Furtado, whose picture illustrates the Figure 3.3, was born in the city of Pombal, Paraíba, and emphasized the State's role in the economy, having integrated the Economic Commission for Latin America and the Caribbean (ECLAC), a United Nations agency, and also created, in 1959, during Juscelino Kubitschek's government, the Superintendence for Development in the Northeast (Sudene).

In João Goulart's government, Furtado was named Minister of Planning for Brazil, and idealized the Triennial Plan for Economic and Social Development, in the planned economy model that had been implanted in the Soviet Union, as a five-year plan, and was later adopted in practically every country.

Furtado, an exponent of developmentalism, looked to establish rules and rigid instruments to control public deficit and hold of inflation. The basis

Figure 3.3   Celso Furtado was one of the greatest Brazilian economists and one of the most illustrious intellectuals of the country. Image adapted from Wikimedia Commons, a collection of free content from the Wikimedia Foundation.

of his methodology, in the decade of 1960, started with the return to the tradition of the Marxist thought, to whom the advancement of technology and the development of the material bases, in certain historic conditions, lead to the adjustment of the other elements. In other words, according to Furtado, alterations in the social structure, or superstructure, are conditioned to modifications in the economic structure, the infrastructure (Medeiros and do Val Cosentino, 2020).

He had his political rights revoked by the military coup of 1964, which deprived the country of some of its greatest intellectuals, and spent a period at the University of Yale's Institute for Development Studies. He later moved to Paris, where he was named by General Charles de Gaulle, as the first foreign tenured Professor for a French university.

For two decades, Celso Furtado was a professor of development Economics and Latin American Economics at the Law and Economic Sciences University at Sorbonne, dedicating himself to teaching and research at the University of Yale, the University of Columbia, and the American University, in the United States, and Cambridge, in England, while also participating in the United Nations.

Furtado, like other philosophers, considers underdevelopment a form of social organization at the core of the capitalist system, instead of a step in the direction of economic development. Terms such as emerging country or in development, only mask the fact the underdevelopment is a specific structural process, and not a phase that the countries considered developed, went through.

The underdeveloped countries had indirect industrialization, or in other words, because of the demand created by industrialized countries. In the specific case of Brazil, there was industrialization reliant on developed countries, and therefore, difficult to overcome without the intervention of the State, to redirect the surplus, that was always allocated to the consumption of the higher classes, to the productive sector (Wikipédia, 2020a).

This did not imply, to Celso Furtado, the necessity of a complete transformation of the productive system, but solely the redirection of the economic and social policy of the country to improve the living conditions of the population.

3.3 Economic Determinism and Wealth and Income Distribution

As observed by Thomas Piketty, in the results of his main study, do not trust economic determinism when it comes to wealth and income distribution, since historically, wealth distribution was always political and depends on how the political, social, and economic agents, perceive the concept of distributive justice (Piketty, 2014).

Furthermore, wealth distribution has a dynamic that can converge or diverge, and there is no natural process to impede the forces that promote inequality. In general, the diffusion of knowledge, the investment in education and qualification lead to convergence, namely, to the reduction of inequality. The law of supply and demand, as well as capital and work mobility typically act in an ambiguous manner but can operate in favor of convergence.

On the other side, acting in favor of divergence, are the processes of accumulation and concentration of wealth, the high remuneration of capital, as well as the salary differentiation, like the one caused by the almost limitless elevation of their own salaries by company executives, or by councils controlled by them.

Moreover, if the return rate of capital, in the form of profit, rent, dividends, and interest rates, for instance, is superior to the annual rate of growth of income and production, for a long period, the chance of divergence in the income distribution elevates. Finally, the scarcity principle, defined by David Ricardo, has the potential to increase prices and contribute to structural divergence.

Piketty summarizes his conclusions in the relation $r > g$, in which r corresponds to the return of capital, and g is the growth rate of the economy, which are obtained from the following equations.

The first one defines the capital return by means of accounting identity,

$$\alpha = r \times \beta, \tag{3.1}$$

in which α corresponds to the participation of capital in the national income and β is the ratio of capital over income.

Furthermore, the ratio between capital and income is given by,

$$\beta = \frac{s}{g}, \tag{3.2}$$

in which s is the savings rate and g corresponds to the growth rate of the economy.

In other words, the inequality $r > g$ is equivalent to $\frac{\alpha}{\beta} > \frac{s}{\beta}$ or, considering that all the variables are positive, $\alpha > g$. This means that the participation of capital in the national income is superior to the income rate of the national savings, what would be, to Piketty, the fundamental force of divergence in the economy (Piketty, 2014).

Marx thought that the main mechanism for the disappearance of the bourgeoisie would be the principle of infinite accumulation, or in other words, the capitalists would accumulate amounts of capital so elevated, that the process would force a decrease in capital income rate, the profit rate (Piketty, 2014).

It can be noted that in case there is no structural growth in the economy, if the growth rate g tends to zero, for a positive savings rate, a conclusion is reached, that can be considered a logical contradiction, similar to the one Marx reached,

$$\lim_{g \to 0} \beta = \lim_{g \to 0} \left[\frac{s}{g} \right] = \infty. \tag{3.3}$$

In other words, capitalists accumulate more and more capital each year, for various reasons, including power hunger or from the high living standard, and that causes the relation of capital to income to increase indefinitely, tending to infinity.

With a relation between capital and income tending to infinity, and revenue from capital r, Given by the Equation 3.1, converges necessarily to zero, as

$$\lim_{\beta \to \infty} r = \lim_{\beta \to \infty} \left[\frac{\alpha}{\beta} \right] = 0, \tag{3.4}$$

making the capital's participation consume all the national income.

This logical contradiction presented by Karl Marx causes a problem, whose solution is the structural growth of the economy, in terms of productivity and population, to balance the process of unrestrained capital accumulation.

The alternative against the decrease in income rate, which has been utilized by certain countries, is to stimulate conflicts, considering that wars destroy goods that need to be replenished.

They destroy property, and unfortunately, lives, to feed gigantic conglomerates of militaristic companies, besides the armies, suppliers, and the entire chain of production, if it is possible to call it that, resulting from bellicose activity (Alencar, 2010).

The United States of America, for instance, learned this procedure of earning profits with conflicts, and generally promote wars in intervals of two and a half years, for over a century (Tannenbaum, 1973).

In the past, they used wars to renew their own reserves, but they had losses. From the teachings of John Maynard Keynes (1883–1946), a British economist that influenced contemporary macroeconomics, things changed.

Keynes defended the economic politics of the intervening State, by which governments should use fiscal and monetary measures to reduce the adverse effects of the capitalism economic cycles, predicted by Karl Marx, which include recession, depression, and growth.

After the Second Great War, Keynes's economic ideas were adopted by the main occidental economic powers, including the United States, and the countries had long periods of growth.

The United States learned, then, that the recession that followed, inexorably, every war, could be avoided with government intervention. From there, they started profiting from every war in which they participated.

3.4 Virtual Companies Exploit the Real Work of Users

The production market was well defined in Adam Smith's time, author of one of the most important economics books of all time, "An Inquiry into the Nature and Causes of the Wealth of Nations," in 1776, that all entrepreneurs should actually read, to understand that Smith was, in reality, a harsh critic of the capitalist's selfishness (Alencar, 2014c).

There were artisans that devised and assembled products for clients, eventually custom- made, as in the case of personal effects, there were farmers, that sowed the land and offered basic necessity products. A simple market, in which it was possible to clearly identify what a product was.

Smith analyzed the regulation process of prices attributed to the market, and noticed that specialization was the key to increase production, as was later verified with the Industrial Revolution.

At the time that Karl Marx published his masterful book "*Das Kapital*," in 1867, that the capitalists did not read and did not like, and that all the socialists should actually read, industrialization was in course in England, products started to be made by machines, in great volumes, and at lower prices, causing unemployment (Marx, 2004).

Marx was the economist that understood the acting forces in the market at the time, who gave history an interpretation distinct from everything that had been published before, when putting economic cycles in the center of the process, and continues to be the most influential of all time.

Nevertheless, the fact is that in the past, there were products that were made by the industrial sector, and sold by commerce. Producers, sellers, and clients were clearly identified.

Few areas, like tailoring, for instance, required interaction between the producer and the client, or direct action from the client in the manufacturing. In time and with the assembly line, however, clothes factories started to dispense the participation of clients in the preparation of clothes.

More recently, restaurants started to exploit the client's work, with self-service. The client started to be part of the productive process.

Gas stations also started to use client's labor, and the owners of the establishments appropriated what they would have to pay attendants. However, they still needed to buy the equipment, like automatic gas pumps and computer systems for payment registration.

With the emergence of the Internet, and the associated services, created by Timothy Berners-Lee, in 1989, and integrated into a platform that he initially called Mesh, and in 1990, renamed World Wide Web (WWW), the sale of products was radically altered.

The main change was not the remote selling of products, a process that existed for centuries, and that always used the postal service for this end, neither the commercialization of computer programs, that has been done since the computer became commercial, in the decade of 1960. Neither was dispensing the salesperson, with a direct link between the developer of programs and the users of the network.

What really changed with electronic commerce was the production relation between the manufacturer and the client. The successful platforms that, in 2014, maintained 87% of the entire American population connected to the Internet, were those in which the users created a considerable part of the final product.

3.5 Virtual Companies Violate Classic Economics

The social networks are products that have been created, mostly, by the clients. They feed them with information, which increments and values the products, invite friends to participate, which is a marketing and advertising process, and even help to develop new facets of the products, which would be the work of the company programmers (Alencar, 2014f).

Furthermore, the users also serve as success and penetration indicators, as privileged informants for market research, and as consumers for the products outsourced by the social networks.

The market value of a social network, such as Facebook, is hard to appraise, but is generally measured in function of the number of users on the platform, associated with the number of accesses produced by these users. Moreover, the users are the ones that create wealth for the information tycoons, working every day to improve their products.

However, since these platforms are virtual and perishable products, a simple prejudiced rumor, like the one that seemed to have affected Orkut, alleging that the platform was only used by maids, that a few years ago was the largest social network in the country, and was shut down in 2014, could destroy Facebook in an instant.

Currently, very little is created in this area, since UNIX was invented based on Multics, which was written in 1965, by a group of programmers, which included Ken Thompson, Dennis Ritchie, Douglas McIlroy and Peter Weiner.

Mark Zuckerberg only copied and improved Orkut's idea, which inherited its name from the chief engineer of the project Büyükkökten Orkut, from Google, who had already copied functionalities from other platforms that used UNIX commands and convinced the users to work for him.

Facebook is simply an application of the mail and talk commands, created for UNIX. Just like Google, it is only an enhanced version of gopher or archie, from the end of the decade of 1980.

Before it, there was Altavista, Bol, UOL, Yahoo, among many others that did more or less the same thing, but today they specialized in other activities or disappeared. Twitter, for instance, a success from a few years ago, is simply the bulletin board system (BBS) application, with files restricted to 140 characters. Which, in turn, is only another application of the mail command from UNIX.

There are still several specialized networks for academia, for researchers and professionals, like Academia.com, ResearchGate, and LinkedIn. Not to mention social discovery networks, like Quora and Twoo. They all appropriate the users' work, providing them a few minutes of fame every day.

It is interesting to note that, following Adam Smith's line of thought, David Ricardo defended the labor theory of value, by which the value of any good or service is determined by the amount of labor incorporated into it. In other words, it is the labor, and not the utility or scarcity, which appropriates value to a good or service when compared to others (Smith, 1986) (Ricardo, 1996).

The most fascinating of this all is that, to Karl Marx's amazement, who also understood that labor was the real source of all value, those entrepreneurs would end up creating a type of factory in which the clients themselves work, several hours a day, just for the pleasure of being there and finding other users, that do the same thing (Marx, 2004).

The social network reflects the model imagined by Adam Smith for an efficient market, in which the producers are very specialized, in themselves, and build their profiles as a production line, over time .

These virtual companies can enjoy the workforce of millions of people, in tens of countries, just by allowing them a space to maximize their utility or happiness, as predicted by James Stuart Mill (1773–1836), the philosopher, economist, Scottish erudite, and father of John Stuart Mill (Balckburn, 1997).

To the despair of classical economists, and violating all the canons of Economics, platforms such as Facebook are virtual monopolies, whose products are available for free.

3.6 How the Money of Private Companies Leave the Country

During the administrations of Fernando Collor de Mello and Fernando Henrique Cardoso, when privatization of state-owned companies was architected, one of the main justifications for selling public patrimony was the investment that foreign companies, the probable buyers, would make in Brazil (Alencar, 2014e).

At the time, Collor de Mello used to say that it was necessary to sell the companies, to impregnate the Country with new technology, which would be brought over by the companies that landed in Brazil. A badly put metaphor, for a Country that had just started to produce their own technology in some strategic areas, such as telecommunications.

But even then, the privatization process started, the companies were sold, in the most part to foreign corporations, or to national consortia which would be denationalized. In both cases, the profit obtained in Brazil would have to be sent abroad.

To this effect, usually, there are some ways of forwarding resources abroad, some legal, others not so much. It is possible to send them via the natural way, Central Bank Cacex, which is a controlled and bureaucratic process. As there are limits to the shipment, it is not possible to send a lot of money, and this ends up not being the preferred route for shipments.

The company can, for instance, acquire overpriced equipment and material abroad, with prices much more elevated than the national equivalent. This is easy to do, but implies the scrapping of the national industry, like in the case of telecommunications, and can be done routinely by companies. The phone companies do not buy telephone exchanges in the Country, since it is better to send the money abroad this way.

Companies can take out loans abroad, and from their headquarters, making it possible to permanently send currency over many years. This procedure is constantly utilized since it allows an elevated and constant flow of resources. In addition, the loan will certainly never be repaid, since there is no interest in that.

After the process is approved, the Central Bank is not worried anymore, and the shipments can last the entire concession period of the public service. Light (Toronto Traction and Light Power), a Canadian company that exploited the energy, trams, communications services, among others, in the Southeast of Brazil during almost a century of concession used this procedure.

In 1990, Brascan, controller of Light, which had only 20 employees working in Canada, profited US$ 20 billion, and all the money came from Brazil. The interesting part was that it was by far the company with the most profit per employee in Canada, with a billion dollars per employee!

There is also a kind of shipment by *courier*, in which employees are used as human mules to transport the money. It is used mostly by software companies, that normally do not have the means to justify the purchase of equipment or another manner of shipment.

Thousands of people travel each year, especially from countries like Argentina and Brazil, taking bags of money to the headquarters. This can involve customs employees, which need to be bribed occasionally.

Usually, those people do not take more than US$ 100,000.00 each trip, since they need to accommodate the money in bags, in a way as to not cause suspicion in the customs. The mules typically use the excuse of going to conferences organized by the companies. It truly is hard to send resources abroad, in a legal manner.

3.7 Corruption is Endemic in the Private

The business world and the public sector have some very different cultures, which ends up complicating the relations between companies and government. Something that is bizarre or a crime in one world, is normal, and might even be the business' *raison d'etre*, in another (Alencar, 2014b).

Profit, the biggest objective of an entrepreneur, is a cursed word in the public circle, and is not well viewed in social organizations of public interest, as well as in religious and scientific societies. A university, for instance, cannot have profit; it needs to invest everything it collects into its maintenance, employee salaries, scholarships for students, and improving its structure.

The creation of companies to accommodate the children and employ family members and friends is ideal for business, industry, or landowners. But it is considered a crime of nepotism in the governmental sector, which can only hire by means of public exams and *curriculum* scrutiny.

Not paying taxes is one of the objectives of almost any entrepreneur, that hires, to this end, dozens of accountants and lawyers, executing several fiscal maneuvers, creating shell companies, changing the corporate name, putting personal assets in the company's name embezzles resources, sells patrimony, to later rent and simulate expenses, an infinity of artifices to cheat taxes. However, this is virtually impossible for state owned companies, that are always carefully and periodically examined by some federal controlling office.

To illustrate the idea of the amount, tax evasion can reach 87% in companies, according to the former executive secretary of the Ministry of Finance, in Brazil. In other words, only 13% of the entrepreneurs pay all of their taxes. The rest use some kind of artifice to avoid paying. As every nation depends on tax collection to pay employees, pensions, and invest in construction work, the tax paying individuals are the ones penalized.

Scheming for profit amongst entrepreneurs is a typical characteristic of the private initiative. This usually does not exist among public entities. It would be strange, for instance, if there was a scheme going on between official banks and universities with profit in mind.

Furthermore, price-fixing among competitors is a common practice but is absent in relations between public companies. As is the formation of cartels in the business environment, with no correspondence in the public environment.

Entrepreneurs pay each other bribes, as a regular activity in private business, to seal deals, make sales, acquisitions, and mergers. Moreover, they call it a commission. The always-present intermediary is not viewed with contempt in private activity and can even be necessary. On the other side, there are no means of paying bribes among public entities, neither is their interest in it, and intermediaries are seen as criminals.

Companies donate millions to political campaigns, directly to the politicians or their representatives, apparently with no business oriented strategic objective. They give money to government or opposition politicians indiscriminately, although they maintain a certain level of discretion in relation to this. When this happens with public companies it is certainly a crime, as has been rumored in several cases around the World.

With so diverse cultures, the association of the private and public sectors is always tortuous and difficult. Entrepreneurs, accustomed to a more liberal, behavioral standard, will certainly band together in cartels, creating shell companies to take part in auctions, setting up schemes of all types to fix prices, and offer commissions to public employees. And some of them will take it, which is a crime.

Corruption is not an isolated fact, a problem in underdeveloped or developing countries, it is not an epidemic. It is endemic to the private sector in the entire World, by the simple nature of capitalist accumulation. When the private sector touches the public sector, is when it starts being considered a crime.

3.8 Tricks for Avoiding Taxes

Tax avoidance is the act of using the possibilities stipulated by law to avoid paying taxes. This is done by companies, banks, industry, and high middle-class citizens (Alencar, 2015d).

It comes as no surprise that the creator of the current banking system, John Law (1671–1729), an economist, before the term was widespread, and son of a rich Scottish family became notable for his life as a gambler, vagrant, womanizer, carouser, and murderer, and spent his life fleeing from justice, in spite of his suggestive name. Figure 3.4 shows Law in his prime, as a finance manager in France.

Law proposed the creation of a centralizing State bank with the monopoly of the financial activities, and defended the use of fiduciary currency, to avoid the flight of residual gold, guaranteeing the survival of the local private banking, in addition to the launch of credit titles backed by revenue from taxes and properties. In other words, Law helped create financial capitalism, with credit multiplication and the broadening of moneymaking opportunities (Gleeson, 1999).

Figure 3.4   John Law, Scottish economist and creator of the current banking system. Image adapted from Wikimedia Commons, a collection of free content from the Wikimedia Foundation.

Among Law's main ideas, there was the theory of value by scarcity, the distinction between exchange value and use value, currency as a means of exchange, the theory that currency offer should not be determined by gold imports or by the balance of trade, in addition to his defense of the introduction of paper currency and titles backed by land and government taxes (de Carvalho, 2012).

Adam Smith classified Law's system, which ended up bankrupting the French *Banque Géneréle*, created for him by the regent of France, in 1716, as "the most extravagant project on banking activities, as well as stock speculation, that the World has ever seen." Karl Marx described Law as "the delightful character that is a combination of a swindler and a prophet" (Gleeson, 1999).

For the act of tax avoidance, tributary planning is used, which is the set of legal systems that aim to lessen the payment of taxes. Fraud and simulation, in the field of tributary law, are illicit forms of tax avoidance and tax evasion.

For instance, it is possible to get internal loans, when a company forges a self-loan operation or to one of its subsidiaries, to avoid paying taxes.

Companies can sell assets and then rent them afterward, to simulate spending and declare losses, instead of profit. The assets are usually sold to the directors or their family members.

Companies can create banks in tax havens to loan money to their subsidiaries. Companies can be created in tax havens in a few hours, with the simple act of renting a post office box.

Another alternative is the payment of *royalties*, when a subsidiary in a tax haven takes on the rights of the brand of a company and starts to charge for its use.

The fiscal orderings that exempt economic operations should usually tax, according to the main principles accepted by the international community, or those that tax at very low rates, typically attract foreign companies and capital, are considered tax havens.

In other words, a tax haven, is the country or autonomous region where the law facilitates the application of foreign capital, offering a type of fiscal favoring, with very low or null tax rates.

The most famous tax havens are: Switzerland, Luxembourg, Hong Kong, United States, Singapore, Jersey Islands, Japan, Germany, Bahrain, Liberia, New Hebrides, Nauru, Uruguay, Monaco, Bahamas, Bermuda Islands, Turks and Caicos Islands, Cayman Islands, Isle of Man, Liechtenstein, Canal Islands and Madeira Islands. But there are over 40 tax havens in the world.

Those territories have flexible financial and corporate legislation in common, they offer rigid financial and professional secrecy, absolute exchange freedom, efficient financial and communications systems, in addition to good social and political stability.

Some institutions, such as the Brazilian Internal Revenue Service, consider tax havens the countries or territories that have a tax rate inferior to 20%. Brazil also classifies as a tax haven, countries whose legislation allows secrecy of corporate composition of companies.

The fact is that companies spend fortunes hiring accountants and lawyers, opening subsidiaries in tax havens, simulating selling assets, hiding profits, among other fiscal tricks, to avoid paying taxes. What seems like insanity is only a game to most entrepreneurs, but a crime against every nation.

4

The Main Concepts in Economics

"The so-called neoliberalism is a totalitarian capitalism."
José Saramago

4.1 Introduction

This chapter presents the main economic concepts and definitions, that are useful to the understanding of the topic. The glossary of Economics available in the appendixes contains a sizeable share of the terminology associated with the area.

4.2 Definitions Associated with Economics

An economic activity represents the set of procedures followed by people to satisfy their needs, by means of production and exchange of goods and services (Smith, 1986).

This activity is regulated by supply, the amount of goods and services that the sellers are willing to commercialize on various price levels and by demand, the amount of goods and services, which can be acquired for a set price, in a given market, during a certain unit of time. Aggregate supply and demand refers to the concept of supply and demand but applied at a macroeconomic scale.

The aggregate supply, also known as total output, is the total provision of goods and services produced within an economy at a given overall price in a given period. The aggregate supply is driven by some factors of production, that include labor, capital goods, natural resources, and entrepreneurship.

A modification in aggregate supply can be attributed to many causes, including changes in the extent and quality of labor, technological innovations, changes in producer taxes and subsidies, an increase in wages or production costs, and changes in inflation rates. Some of these factors lead to positive changes in aggregate supply, while others render aggregate supply to decline.

Some factors potentially increase the aggregate supply, including population growth, rising physical capital stock, and technological progress. On the other hand, wage increases place downward pressure on aggregate supply by increasing production costs.

The annual sum of investments, net exports, and consumption expenditures of the population and government in a country is called aggregate demand. The components of aggregate demand are consumer spending, business spending, government spending, and exports minus imports.

Aggregate demand also refers to the demand for the country's GDP, and the measure of demand for goods and services at all price levels. A price level is the hypothetical overall price of goods and services in the economy. It is determined using the Consumer Price Index (CPI), which is a measure of the weighted price of a basket of typically purchased goods and services in the economy.

The aggregate demand of an economy is the sum of all individual demand curves from different sectors of the economy. It is typically the sum of four components: consumption spending, government spending, investment spending, and net exports.

Consumption spending is the largest component of an economy's aggregate demand, and it has to do with the total spending of individuals and households on goods and services in the economy. Consumption spending depends on several factors, such as disposable income, *per capita income*, overall debt, consumer expectations of future economic conditions, and market interest rates.

Government spending is the total amount of expenditure by the government on infrastructure, investments, defense, and military equipment, public sector facilities, healthcare services, and government employees. It excludes the spending on transfer payments, such as pension plans, subsidies, and aid transfers to other countries that are in need.

An important point to note is that consumption spending does not include spending on residential structures, which is accounted for in the investment spending component.

Investment spending is the total expenditure on new capital goods and services such as machinery, equipment, changes in inventories, investments in nonresidential structures, and residential structures. Investment spending depends on factors such as interest rates, future expectations regarding the economy, and government incentives, which include tax benefits and subsidies.

Net exports represent the difference between exports and imports. Exports are products that are manufactured by domestic producers and sold abroad, while imports are products that are made abroad and imported for domestic purchase.

Remember that aggregate demand is the total demand for domestically produced goods and services. Therefore, exports are added to the aggregate demand, whereas imports are subtracted. The measure of exports minus imports is the net exports, which is important to determine the aggregate demand.

Economics is the Science that studies the economic laws to be followed, to maintain a high level of productivity, to improve the standard of living of the populations, with the correct usage of resources. Socialism is the doctrine that advocates the organization of an egalitarian society, free from exploitative relations between social classes, and that ensures the primacy of collective interest over individuals.

An economic system based on private ownership of the means of production and their use for profit. Being a regime based on the dissociation between the owners of the means of production. These means are used for the purpose of profit. Employers carry out production by paying a wage, which remunerates their workforce (Heilbroner and Thurow, 1982).

By analogy, a capitalist is an individual who temporarily invests in emerging companies with evident growth potential, and whose objective is to obtain profitability above the alternatives available in the financial market. In general, this individual participates in the management of the company for the duration of the investment.

A subdivision of Capitalism, Neoliberalism is the political-economic doctrine that constitutes an adaptation of the principles of liberalism, in which individual activities and initiatives are encouraged as opposed to those of the State.

According to Luiz Carlos Bresser-Pereira (1934–), a Brazilian economist, political scientist, social scientist, business manager, and lawyer, the liberals preach that the State is incapable. However, this does not correspond to reality, as history shows.

The State has played a fundamental role in all of the episodes of industrialization, and all the cases of industrial revolution occurred in the scope of developmental States, starting with the Industrial Revolution in England, still within the context of mercantilism, which was the first historic form of developmentalism (Bresser-Pereira, 2019).

In the 1990s, the neoliberal policies and reforms did not fulfill what was promised. The countries submitted to neoliberal order incurred current account deficits, produced by exchange rate populism, and resulted in low growth rates, financial instability and an increase in inequality.

However, at the beginning of the 2000s, in view of the incapability of neoliberal reforms promoting growth in the countries, developmentalism, which had been proposed in Brazil by Celso Furtado, emerged once again, as a historic phenomenon, as a strategy and way of coordinating capitalism and as a new theory of economic development.

Free trade was initially described as an export and import activity by Josiah Child (1630–1699), an English economist, merchant, and politician, and one of the proponents of mercantilism, in 1668.

The study of classical economics began in 1817, David Ricardo laid the foundations for free trade and the specialization of work. In 1948, John Stuart Mill created the basis of liberal economics, defending commerce and social justice.

An economy can be open, one in which there is freedom for imports and mobility of factors across geographical borders, or centralized, in which the government is responsible for the majority of the economic decisions.

The formal economy is the part of an economic system that respects the payment of taxes and the registration of employees and transactions, while the informal economy is the part of an economic system, consisting of small groups of production, sales, or services, which does not respect the payment of taxes and the registration of employees and transactions.

The globalized economy is the economic situation in which barriers to trade and the flow of capital between countries are eliminated. More recently, information economics describes markets where a buyer has better information than another, according to George Akerlof.

4.2.1 Macroeconomics

A branch of economics that studies, on a global scale and by statistical and mathematical means, economic phenomena, and their taxation in a structure or in a sector, verifying the relationships between elements such as national income, the level of prices, the interest rate, the level of savings and investments, the balance of payments, the exchange rate, and the level of unemployment. The first analysis of an entire economy was carried out by François Quesnay, in 1758.

4.2.2 Microeconomics

Microeconomics is the theoretical process designed to determine the general equilibrium conditions of the economy based on the behavior of individual economic agents, producers, and consumers. It deals with the way in which the individual entities that make up the economy, private consumers, commercial companies, workers, and large landowners, producers of private goods or services act reciprocally (Chacholiades, 1986).

4.2.3 Econometrics

Econometrics is the strand of Economics that uses a set of statistical tools, mainly estimation and hypothesis testing, to model and understand the relationships between economic variables. It combines economic theory, mathematical economics, computer science, statistics and stochastic processes, with the purpose of providing numeric values to economic relations, such as marginal values, averages, and trends, to verify economic theories (Koutsoyannis, 1977).

4.3 Activities of Economic Nature

Economic development is the process that increases the GDP and the following increase to the standard of living of the population, measured by income *per capita*, obtained dividing national income by population.

The standard of living depends on the national income, which is obtained subtracting the capital depreciation from the GDP, to obtain the net domestic product (NDP), and adding to this result the net income received from abroad, or subtracting the net income sent abroad, according to the country's situation.

Therefore, the national income calculated in year n, $R_N(n)$, is the sum of the internal production, $P_I(n)$, in the same year, with the income received from abroad for that year,

$$R_N(n) = P_I(n) + R_E(n). \tag{4.1}$$

In this context, the national wealth, also called national capital, represents the entire value of that which the population and the government possess, at market value, at a certain time, which can be commercialized in some market (Piketty, 2014).

Furthermore, the production and income, in a certain year, be written as the composition of income from the capital plus the income from work,

$$R_N(n) = R_C(n) + R_T(n). \tag{4.2}$$

The World income equates to the production of the entire World, i.e. the income received and the income sent abroad is balanced through accounting on a global scale. This is a conceptual and accountable fact, which makes it impossible that the income exceeds the amount of wealth produced, in a year, unless the country is in debt. However, this cannot occur, of course, on a global level (Piketty, 2014).

The economic statistics of the countries are elaborated and published by governmental agencies, usually together with the central banks of those countries. Those studies contemplate the stocks of actives and passives, in addition to series of income flux and production.

Eventually, a decrease in job availability, production, and demand for national products occur, characterizing an economic slowdown, which can retract the aggregate demand of a country's economy, generating an economic cooldown.

The decreasing production and consumption of goods and services produced in the country can cause a negative variation in the prices of the economy, generating deflation, which can evolve into depression, that phase of economic cycle, characteristic of capitalist economies, marked by the decrease in production, a downward trend in prices, and the increase in unemployment.

The Wall Street stock market crash, in 1929, represented a drastic decrease in the heat of securities and shares negotiated, and initiated the Great Depression in the United States.

4.4 Economic Conditions

The ideal State as conceived by Plato, around 380 BC, is that in which property belongs to all, and work is qualified. To Plato, a fair State should allow the citizens to develop their capacities, talents, and interests, above all in relation to the practice of virtue and justice, as prescribed in The Republic (Plato, 1970).

The Economy functions like a living organism since it is the result of the interactions of innumerable physical and legal agents. Those actions can lead to growth, with improvements to the financial health of the State; conversely, it can lead to stagnation and the sickening of the Economy.

The economic stability indicates the maintenance of full employment, a general stability in prices and equilibrium in the balance of international payments.

In general, this comes with monetary stability, the balance of the currency value reached in a country by means of the control of the monetary supply, interest rates and public debt.

Economic instability is the economic situation in which fluctuations are observed in the level of production, employment, or consumption. This can be followed by inflation, which is a general phenomenon of adjustment, through monetary means, of the existing tensions in a socioeconomic group, which is characterized by the increase in the general price level and the depreciation of the currency.

The inflation index reflects the percentage increase in prices during a certain period. It consists of linking the value of a capital or an income to the evolution of a reference variable, such as price, production, productivity, for example.

A recession is a decrease in economic activity, with a fall in production and an increase in unemployment, which occurs when the volume of wealth that a country produces (GDP) decreases in relation to what it produced in the previous year.

Stagnation is a factor resulting from the demand for investment, exports, consumption, or economic activity in general, with an impact on production. Stagflation is the economic situation characterized by the conjunction of a tenet to stagnation or recession, accompanied by inflation. Hyperinflation is the consistent increase in prices in the economy, or general lack of price control.

Income distribution is the process by which income, from profit, wages and other income is divided between regions, companies, or groups of people, it is one of the objectives of any government, not necessarily for a leader's altruistic reasons, but because it maximizes economic growth and elevates common welfare.

The Gini index has been used to evaluate how wealth is shared in the population. This index was developed by sociologist and statistician Corrado Gini, as a means of measuring the division of wealth in societies. It measures the distribution of income (in some cases consumption expenditures) between individuals or households, within a deviation from a perfect distribution. Its value varies from 0 (or 0%), indicating perfect equality, to 1 (or 100%), which indicates perfect inequality.

4.5 Economic Sectors

The economic sector is the part of the economy, represented by private companies, whose decisions are determined by the market. It is responsible for the production of goods and services in general.

It is natural that the economic sector usually presents seasonality, a denomination of the period of the year with the highest activity in a given sector of the economy. The industry has a higher level of activity in the months of September and October, when production increases to meet the orders of the trade for Christmas sales.

The service sector performs tasks or offers assistance that contributes to the satisfaction of individual or collective needs, in a way that does not transfer the property of material goods.

An economic system is a set of related legal and social institutions, in which certain technical means are employed, organized according to certain dominant causes, to ensure the realization of the economic balance.

The non-financial public sector is composed of federal, state, and municipal public companies, with the exception of banks, title distributors, brokers and other companies with permission to operate in the financial market.

The banking system is the set of private and state financial institutions that offer services such as custody, loan, and investment of money.

Public debt is everything the government spends on loans and securities issues. Lastly, securitized debt is the title of responsibility of the National Treasury issued as a result of the assumption and renegotiation of the Union's debts or assumed by it under the law.

4.6 Utility and Value

The utility is the quality of what is used by the economic agent. Utility is one of the basic notions of the economy, such as that of value.

Jeremy Bentham (1748–1832), philosopher, jurist, and one of the last illuminists to propose a moral philosophy system. Utilitarianism represents a theory formulated by Jeremy Bentham, in 1791, whose aim is to provide the greatest happiness for the greatest number of people.

The available values form the set of the company's liquidity or credit securities that can be quickly converted into currency. Market value refers to the sum of all the shares of a company, which is different from the book value, as it does not take into account factors such as the net debt and the assets of the corporation.

Nominal value is the share price, mentioned in a company's registration letter. It is an estimate that is performed to define the value of an asset according to the government's view.

4.7 Productive Economic Activity

The general level of economic activity in a capitalist system, according to John Maynard Keynes (1883–1946), is determined by the disposition of the entrepreneurs' in investing capital. Economic hardships arise when this interest is blocked by problems, making it difficult or impeding the accumulation of capital to invest (Heilbroner and Thurow, 1982) (Roncayolo, 1973).

The detection of these problems, in time to propose alternatives or stimulus to economic activity, is important and goes through the formulation of mathematical models that predict the behavior of stochastic variables, such as the rate of companies opening in a certain period of time, the number of companies in a certain moment in the economy, or the rate of bankruptcies per year. However, there are not enough studies published about the use of probability theory and stochastic processes tools for the prediction of these economic variables (Alencar, 2009).

Augustin Cournot (1801–1877), a French philosopher, mathematician and economist, started the area of research in Mathematical Economics, with the publishing of the book *Researches sur les Principles Mathématiques de la Théorie des Richesses*, in 1838. The objective of the discipline is to apply mathematical methods to represent economic theories and analyze problems proposed by Economics, with the goal of formulating or deriving theoretical relations in a generic, rigorous, and simple manner (Cournot, 1838).

Cournot's seminal book was ignored by the economists of the time, just as the author had predicted in the preface of the work. Incredibly, Cournot went even further, and linked the possible reasons for the eventual prejudice that the work would suffer from the time's philosophers: the false point of view from which his theory was looked at by the small number of people that thought of applying mathematics to it; and the false notion formed about that analysis by those well versed in Political Economics, to which the mathematical sciences are not very familiar. He observed that the economists of the time had a certain prejudice against the use of mathematical formulation, reinforced by philosophers such as Adam Smith.

Eventually, the book was discovered, in 1880's, by the founders of the marginal focus in Economics, William Stanley Jevons (1835–1882), British, one of the founders of Neoclassic Economics and formulator of the Theory of Marginal utility, Léon Walras (1834–1910), French economist and mathematician, creator of the General Equilibrium Theory, and Alfred Marshall (1842–1924), one of the most influential British economists, whose book, Principles of Economics gathered the theories of supply and demand, of marginal utility, and of production costs.

Cournot's method made him the precursor to Mathematical Economics, especially in the marginal area, with differential calculus applied to economic analysis. His conceptions were, initially contested by Joseph Bertrand (1822–1900), a French mathematician, historian of sciences and academic, and by Francis Ysidro Edgeworth (1845–1926), a British economist of liberal inspiration (Fischer, 1898).

Cournot's focus was redeemed by the works of the American mathematician John Forbes Nash (1928–2015), about game theory, differential geometry and partial differential equations, for the analysis of the forms of competition, and presently constitutes one of the fundamental models of Industrial Economics and the Theory of Imperfect Competition.

Cournot developed the first formula for the rule of supply and demand, as a function of the price of the merchandise, and was the first to draw the supply and demand curves on a graph, anticipating the work of Alfred Marshall in three decades. Furthermore, Cournot introduced the ideas of function and probability in economic analysis developed prevailing theories on monopolies and duopolies, competition, and introduced the concept of strategic equilibrium.

Studies on the analysis of the composition of the growth of services in the Brazilian economy were accomplished, to test the hypothesis that intermediary services tend to elevate their participation in the sector, be it by the emergence of new products, be it by the implementation of organizational innovations, be it by the escape attempt of excessive costs with work.

In the Brazilian case, the hypothesis was not confirmed as a result of the low growth rates of the economy in the period evaluated, which influenced the composition of the growth of services (Rocha, 1999).

Models extracted from the queuing theory have been used to study, for instance, the differences in the rates of growth of countries, as well as changes in the growth process through time, with the help of Markov chains, named after Russian mathematician Andrei Andreyevich Markov (1856-1922), in which transition probability matrices are attributed to certain countries (Norris, 1997).

Parameters of the matrix, variants in time, were estimated conditioned in relation to the quality of the institutions and the level of investment (Morier and Teles, 2011).

The effects of the economy on fertility and mortality were central in the analysis of Thomas Malthus (1766–1834). Other economists such as Adam Smith, Joseph Schumpeter (1883–1950) and David Hume (1711–1776) also discussed the connection between population and subsistence resources.

According to his predictions, thriving prosperity would lead to a larger increment in population growth, until the limit of food supply was reached. In spite of the results of empirical studies not having validated Malthus's paradigm, the population started being treated as an endogenous element derived from social and economic conditions (Currais, 2000).

In the analysis of the evolution of the institutions, a theoretical impossibility is perceived when starting from a natural state, free of institutions, to reach the emergence of institutions. It is suggested that the emergence of institutions in the real world is assisted by the development of concordant habits, particularly as a result of emergent channels and restrictions (Hodgson, 2001).

The evaluation of the results of public spending on distinct levels of government on economic growth indicate that the investment must be prioritized, especially by the state realm, in detriment of expenses with consumption, subsidies and transfers, since the public sector is not very efficient in dealing with these expenses. Keynes continues as a model when thinking of public investment and the investment to facilitate the opening and consolidation of companies is important, and depends on behavioral models of the economy in this sector.

The expenses with consumption, subsidies, and transfers, in relation to the GDP, for the consolidated government, that between 1950 and 1980 was on average 18% of the GDP, having increased to 26% in the two following decades, while the investment had its participation in the GDP reduced to almost 30%, in the same periods. However, the public expenses can promote economic development, especially with the increase of the attributions of the states in relation to the Union, and the investment in detriment of expenses with consumption, subsidies and transfers (Rodrigues and Teixeira, 2010).

Governmental stimulus to the economy has occurred, in a planned fashion, since the end of the XIX century in the country. In the analysis of the process of industrial formation in Brazil, from the first manufactures implanted in parallel to the agro-export model of the last years of the XIX century and the first years of the XX century, considering the exhaustion of this model, it could be verified that industrialization substituted imports, at first in a non-induced fashion, and later with planning oriented and commanded by the government (Mattei and dos Santos Júnior, 2009).

The economic structures developed by the economists through the 1990's and the start of the millennium, directed their Works on game theory, specifically the stability of the strategic equilibrium and signalized games of stable equilibrium.

Theories of economic growth and business cycles lead researchers to study the global and local asymptotic stability of optimum control systems applied to the theory of economic growth, the search for the solution to dynamic Hamiltonian systems of optimum economic growth, the relationship between balanced growth and intertemporal efficiency in the accumulation of capital, and the description of the structure and stability of competitive dynamic systems.

Depending on the market structure being studied and associated with hypotheses of the behavior of the individual economic agents, diverse results appear for the competitive equilibrium stability (Farias, 2010).

David Ricardo, who was in favor of international liberalism, elaborated the theory of competitive advantages, which became the basis for the theory of international commerce. The logic scheme elaborated by Ricardo consolidated the defense of the system of global commerce anchored in the gold standard and free trade, which only came into disuse in 1944, with the Bretton Woods agreement (Ricardo, 1996).

An econometric study demonstrated, under the hypothesis of Joseph Schumpeter, that the trend to innovate grows with the size of the company for a specific set of companies that interact with universities and research institutes. The results suggest that big companies tend to be more innovative than micro and small companies when it comes to product innovations. However, this size advantage is not verified when dealing with process innovations (Póvoa and Monsueto, 2011).

Large companies have a propensity to enjoy a longer average lifespan than small companies do, this might be the result of interactions with universities and larger investments in innovation that result in better products. The study of the increase in number and vigor of the companies can direct the government in the elaboration of public policies for the sector, which stimulate the association between public universities and the private initiative.

In the opening of a company, two antagonistic forces can be identified. On one side, the prospect of obtaining monopolistic profits is an incentive. On the other side, the risk adverse agents are worried about avoiding the income insecurity associated with self-employment. A small number of firms in the intermediary goods sector characterizes market equilibrium. This has implications for the long-term growth of the economy, which is a function of the portion of entrepreneurs in the population.

The risk of self-employment is also expressed in the high failure rates of enterprises. According to Panel Study of Income Dynamics (PSID) of the United States, the exit rates in the first year, in 2000, were 35%. The entrepreneurs' income is a lot more volatile than wage earner's income (Clemens and Heinemann, 2006).

Curiously, studies seem to indicate that an affluent economy might not provide the adequate incentive for low-income economic agents, in a supply and demand analysis (Chowdhury, 2013).

A study on the cycles of growth and business in the United States, accomplished using a multivariate qualitative Hidden Markov Model (HMM), concluded that markets rarely fail in detecting economic inflection points, with a minimum time of three to six months. Which is an incentive to adopting queuing theory models (Bellone and Gautier, 2004).

Despite the direct relation of macro and microeconomic factors with the quota, as well as company dynamics, there are few studies proposing probabilistic models for the behavior of companies' stock in an economy.

5

Basic Mathematical Analysis

"In regards to the price of commodities, the rise of wages operates as simple interest does, the rise of profit operates like compound interest."
Adam Smith

5.1 Basic Concepts in Economy

The demanded and supplied quantities of a given commodity are functions of several variables, that include the commodity price, the prices of equivalent goods, the amount of money available for the purchase, interests involved, and other possible reasons. Most of those variables are, in fact, stochastic processes, and the complete analysis would involve statistical correlation, covariance analysis or, perhaps, stochastic differential equations.

In basic economic analysis, however, the demand and supply are considered to be dependent only on the commodity prices, which is considered the most important variable. The demand and supply curves are usually not linear but, for the range of interest, they are, as a first approach, considered approximately linear (Weber, 1982).

Figure 5.1 depicts an example of the curves of supply (S) and demand (D). Note that the curves are, usually, shown in the first quadrant, considering that prices (P) and quantities (Q) are, usually, either positive or zero.

Regarding basic economic analysis, a negative price would imply the payment to buyers for the removal of goods from the market. On the other hand, a negative supply indicates that the goods are absent from the market, either because they are not produced, or because they are withheld by the seller, who waits for a better price.

A negative demand denotes a very high price for the goods, which paralyzes the market activity until the product is offered at a satisfactory price. The occurrence of these cases is uncommon, and they are left for more advanced texts.

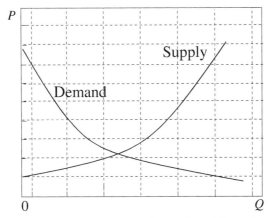

Figure 5.1   Graph that illustrates the supply and demand curves.

5.2 Demand Curves

The following demand curves indicate usual and pathological cases for idealized economic markets. The graphs are linearized to emphasize the important characteristics of each case. They can be used as references for the analysis of real cases.

The negative demand curve is the usual case in a healthy economy. This means that the price increases as the quantity demanded decreases. On the other hand, if the price decreases the quantity demanded increases. Figure 5.2 shows an example of negative demand.

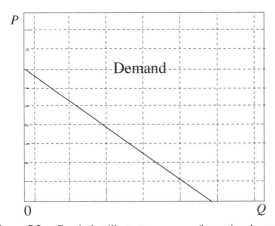

Figure 5.2   Graph that illustrates a curve of negative demand.

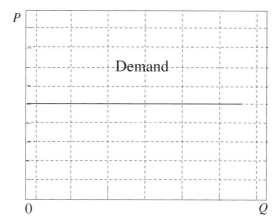

Figure 5.3   Graph that illustrates a curve with slope of demand zero.

If the slope of a demand curve is zero it indicates a constant price regardless of demand. This can happen, for instance, when the prices are kept constant because of a government decree to freeze prices to curb inflation. Figure 5.3 presents an example in which the slope of demand is zero.

The preceding example also illustrates the case of a perfectly elastic demand curve, in which elasticity of a function that relates the quantity demanded and the price, $Q = f(P)$, at a specific point, or point elasticity η, is defined as

$$\eta = \lim_{\Delta P \to 0} \frac{\Delta Q/Q}{\Delta P/P} = \lim_{\Delta P \to 0} \frac{P}{Q} \times \frac{\Delta Q}{\Delta P} = \frac{P}{Q} \times \frac{dQ}{dP}, \tag{5.1}$$

in which ΔQ and ΔP represent the absolute changes in quantity and price. The relative, or percentage, changes in quantity and price are $\frac{\Delta Q}{Q}$ and indicated by $\frac{\Delta P}{P}$.

Figure 5.4 shows an example of undefined demand, indicating a constant demand regardless of price. This is usually the case for military expenditures. This is an example of perfectly inelastic demand.

Elasticity can be used to analyze the responsiveness of demand for a commodity to changes in its price. Because the slope of the demand curve is negative, because the price and quantity normally move in opposite directions, which means that the first derivative is negative, then the elasticity is usually negative $\eta \leq 0$.

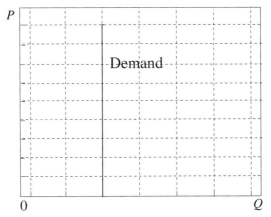

Figure 5.4   Graph that illustrates a curve of undefined demand.

The responsiveness of different commodity demands to price changes can be compared, considering that the elasticity is dimensionless. The demand is usually classified into the following general categories:

- Perfectly elastic, when $\eta = -\infty$,
- Perfectly inelastic, when $\eta = 0$,
- Relatively inelastic, when $-1 < \eta < 0$,
- Relatively elastic, when $\eta < -1$,
- Unit elastic, when $\eta = -1$.

Therefore, the demand is elastic if $|\eta| > 1$, unit elastic if $|\eta| = 1$, and inelastic if $|\eta| < 1$. The price elasticity of demand is of interest to a company. For instance, if the demand is unit elastic, the total revenue is unchanged by a decrease in price. When the demand is inelastic, the total revenue is decreased by an increase in price.

In order to summarize the concepts related to economic elasticity (Kreinin, 1987):

- The demand for a good is elastic when a change in price has a large effect on the quantity of the demanded good. This means that a decline in price yields an increase in the value of purchases.
- The demand for a good is inelastic when a change in price has a small effect on the quantity of the demanded good. In this case, a price reduction produces an increase in the purchased quantity that is less than proportional.

- The demand for a good is unit elastic when the relative change in quantity demanded is equal to the relative change in price. In other words, any given percentage change in price exactly equals the resulting percentage change in quantity purchased.

5.3 Supply Curves

Regarding supply, the following curves indicate usual and pathological cases for idealized economic markets. They are instructive as references for the analysis of real cases. The slope of the supply curve is usually positive, indicating that the quantity supplied by the market increases as the price increases, and diminishes when the price rises. Figure 5.5 shows an example of the positive supply.

If the slope of a supply curve is zero it indicates a constant price regardless of supply. This is the usual case with contracts issued by telephone companies. Figure 5.6 presents an example of a zero supply curve.

Figure 5.7 shows an example of an undefined supply, indicating a constant supply regardless of price. This can happen, for example, when a company reaches its production limit.

5.4 Average and Marginal Costs

In economics, fixed costs are expenditures that do not change regardless of the level of production, at least not in the short term. For example, the rent on

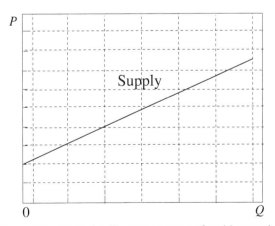

Figure 5.5 Graph that illustrates a curve of positive supply.

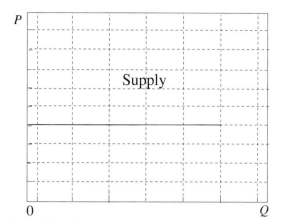

Figure 5.6   Graph that illustrates a curve of zero supply.

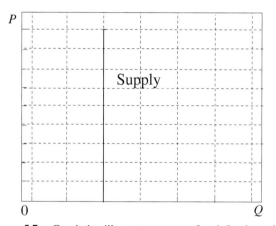

Figure 5.7   Graph that illustrates a curve of undefined supply.

a factory or a retail space. Once the lease is signed, the rent remains the same independent of the production level, until the end of the lease.

Fixed costs include, for example, the cost of machinery or equipment to manufacture the product, research and development costs to develop new products, advertising expenses to popularize a brand name. The level of fixed costs varies according to the specific line of business.

For instance, manufacturing computer chips demands an expensive factory and requires costly equipment. On the other hand, an advertising and marketing agency, in the beginning, might only demand a place to allocate people and a few computers.

Variable costs are incurred in the act of producing. An increase in production implies a rise in the variable cost. Labor is treated as a variable cost since producing more goods or services demands more workers or more work hours. Variable costs may also include commissions and raw materials.

The Total Cost (TC) of a product is the total economic cost of production. It is composed of variable, fixed, and opportunity costs. The fixed costs include the accounting expenses, which do not change based on the output, for example, the building costs, insurance, and property taxes.

The variable costs are the accounting expenses that change based on the output, for instance, the number of goods produced, the number of employees and packaging costs.

The opportunity costs include all of the other expenses that are incurred, that are not usually accounted for. As an example, the amount of interest that could be obtained from another investment.

The Average Cost (AC), also called average total cost, is the total cost divided by the quantity produced. The Marginal Cost (MC) is the incremental cost of the last unit produced.

Suppose that the total cost $C(x)$ of producing and marketing x units of a commodity is given by the function

$$C(x) = f(x). \tag{5.2}$$

Then, the average cost per unit is

$$\frac{C(x)}{x} = \frac{f(x)}{x}. \tag{5.3}$$

If the output is increase by an amount Δx, from a certain level x, there is a corresponding increase in cost of $\Delta C(x)$. Therefore, the average increase in cost per unit increase is given by

$$\frac{\Delta C(x)}{\Delta x} = \frac{\Delta f(x)}{\Delta x}. \tag{5.4}$$

If one takes the limit, as the increment $\Delta x \to 0$, one obtains the marginal cost $c(x)$, defined as

$$c(x) = \lim_{\Delta x \to 0} \frac{\Delta C(x)}{\Delta x} = \lim_{\Delta x \to 0} \frac{\Delta f(x)}{\Delta x} = f'(x). \tag{5.5}$$

The marginal cost is the derivative of the total cost function, with respect to x, that is, the rate of increase in total cost with an increase in output. It is the change in the total cost that arises when the quantity produced is incremented by one unit, that is, it represents the cost of producing one more unit of a good.

5.5 Market Equilibrium

When the demanded quantity of a commodity equals the supplied quantity, the market is said to have reached the point, or price, of equilibrium. The equilibrium price and the equilibrium quantity are found at the point of intersection of the supply and demand curves. Mathematically speaking, the simultaneous solution of the supply and demand equations provides the equilibrium price and quantity.

For the solution to be meaningful, the obtained values for the price and quantity must be positive or zero, that is, the demand and supply curves must intersect in the first quadrant.

Figure 5.8 shows a graph that illustrates a meaningful market equilibrium.

5.6 Simple and Compound Interests

The interest rate is the amount a lender charges for the use of assets, including cash, consumer goods, or assets, expressed as a percentage of the principal. The interest rate is typically noted on an annual basis known as the Annual Percentage Rate (APR).

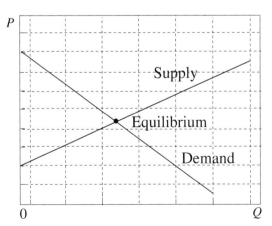

Figure 5.8   Graph that illustrates a meaningful market equilibrium.

Simple interest is a simple method of calculating the interest charge on a loan. The amount obtained by the simple interest $I(n)$ is determined by multiplying the principal P by the daily interest r rate by the number of days n that elapse between payments,

$$I(n) = P \times r \times n. \tag{5.6}$$

Compound interest is the addition of interest to the principal sum of a loan or deposit. It is the result of reinvesting interest, rather than paying it back. Therefore, the amount of interest in the next period is earned on the principal sum plus any previously accumulated interest.

For an interest rate of $100i$ per year that is compounded (payable) k times per year, after n years the original amount of money P, called the principal, becomes

$$I(n) = P \left(1 + \frac{i}{k} \right)^{nk}. \tag{5.7}$$

The equation for compound interest has four variables, aside from the amount accumulated $I(n)$.

If the number of installments per year k is large, that is, if the interest is accrued continuously, in the limit one obtains

$$\lim_{k \to \infty} I(n) = \lim_{k \to \infty} P \left(1 + \frac{i}{k} \right)^{nk} = P e^{in}, \tag{5.8}$$

in which $e = 2.718$ is the base of natural logarithms. Therefore, the debt can grow exponentially, in this case.

Mortgages typically use simple interest. However, some loans use compound interest, which is applied to the principal but also to the accumulated interest of previous periods. As a general rule, a loan that is considered low risk by the lender is charged a lower interest rate. On the other hand, a loan that is considered high risk will have a higher interest rate.

Individuals borrow money to purchase homes, cars, fund projects, launch or fund businesses, or pay for college tuition, for example. Businesses take loans to obtain working capital, fund projects, and expand operations, for instance, by purchasing fixed and long-term assets such as land, equipment, buildings, and machinery. The borrowed money is usually repaid either in a lump sum, at a pre-determined date, or in periodic installments.

5.7 Income Distribution

Vilfredo Fritz Pareto (1848–1923), an Italian civil engineer, sociologist, economist, political scientist, and philosopher, made important contributions to economics, mainly in the study of the income distribution. Figure 5.9 shows a portrait of Vilfredo Pareto.

He proposed the Law of Income Distribution, which states that the number of individuals N from a given population of size A, whose income exceeds x, is given by (Pareto, 1896) (Pareto, 1897)

$$N(x) = \frac{A}{x^m}, \qquad (5.9)$$

in which m is a constant to be found.

Figure 5.10 shows a plot of Pareto's formula, for the number of incomes $N(x)$ that exceeds x. For incomes that are above the subsistence level,

Figure 5.9   Vilfredo Fritz Pareto, in a photo by an unknown author. Image adapted from Wikimedia Commons, Public Domain, https://commons.wikimedia.org/w/index.php?curid=6 8615041.

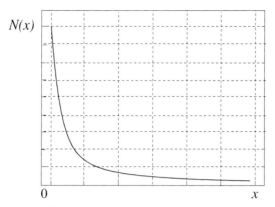

Figure 5.10 Plot of Pareto's formula, for the number of incomes exceeding x.

statistical data indicate that Pareto's Law is generally fairly accurate (Weber, 1982).

The previous equation, which has proved remarkably close to the observed data, is usually put in the following form, by taking the logarithms of both sides,

$$\log N(x) = \log A + m \log x. \tag{5.10}$$

Pareto argued that the distribution of income and wealth, in every country, is highly skewed, with few people holding most of the available wealth. He created modern microeconomics and also questioned the utilitarian philosophy. The Pareto index is a measure of the inequality of income or wealth distribution.

Pareto originally used the formula he proposed to describe the allocation of wealth among individuals since it seemed to show quite well the way in which the major portion of the wealth of any society is owned by a smaller percentage of the people in that society. He also used it to describe the distribution of income in that society.

The Pareto distribution is used today in many areas of research, including: the sizes of human settlements, file size distribution of Internet traffic which uses the TCP protocol (Castro et al., 2013), hard disk drive error rates, the values of oil reserves in oil fields, the length distribution in jobs assigned to computers, the standardized price returns on individual stocks, the sizes of sand particles, the size of meteorites, the severity of large casualty losses for certain lines of business, such as general liability, commercial auto, and workers compensation (Wikipedia contributors, 2021).

5.8 Consumption and Savings

The consumption function is the relationship between disposable income and total consumption. It is usually assumed that an increase in income expands consumption, and a decrease in income diminishes consumption. But there is a difference between the increase in income and the rise in consumption, that is called the marginal propensity to consume, or the rate of change in consumption as disposable income changes, which is greater than zero, but less than one (Weber, 1982).

If the consumption, or utilization, function is given by

$$u = g(e), \tag{5.11}$$

in which u is the national consumption, or utilization, and e is the national income, or earnings, both given in the same units, the the marginal propensity to consume is

$$\frac{du}{de} = g'(e). \tag{5.12}$$

Because the disposable income equals consumption plus savings, one can write

$$e = u + s, \tag{5.13}$$

then, the marginal propensity to save is obtained taking the derivative of both sides of the equation, to give

$$\frac{ds}{de} = 1 - \frac{du}{de}. \tag{5.14}$$

Investment is considered as capital formation in national income analysis, and represents an increase in real capital. Investment and consumption are related, and an initial investment spending may result in an increase in revenue of several times that amount.

The ratio of the resulting increase in income to the rise in investment is called the spending multiplier k, which is related to the marginal propensity to save or to consume, and is given by

$$k = \left[\frac{ds}{de}\right]^{-1} = \frac{1}{1 - \frac{du}{de}}. \tag{5.15}$$

The factor k, also called the Keynesian multiplier, was proposed by the economist John Maynard Keynes, in 1936, in his book "The General Theory of Employment, Interest, and Money," a classical work in economics.

One should note that if $\frac{du}{de} = 0$, then $k = 1$, that is, if none of the additional income is spent, the total increase in revenue equals the initial expenditure. If $\frac{du}{de} \to 1$, then $k \to \infty$, that is, if all the additional income is spent, the total rise in revenue goes to infinity.

In theory, a rise in private consumption spending, investment expenditure, or net government spending, which means gross government spending minus government tax revenue raises the total GDP by more than the amount of the increase.

Keynes' main idea was that recessions and depressions could occur because of inadequate demand in the market for goods and services. In response to widespread unemployment and low levels of economic activity in the world. Therefore, Keynes recommended an increase in government spending to encourage demand for goods and services in the market. The reasoning was against the classical economic policy of *laissez-faire* and minimal government interference.

In practice, the multiplier applies to any type of expenditure, and it applies when spending decreases, as well as, when it increases. If business confidence declines and investment decreases, the change can reduce aggregate expenditures, and then induce an even larger effect on real GDP, because of the Keynes multiplier effect.

5.9 Model for the Future Price of Stock

The Lognormal distribution, presented in Appendix A, typically appears as a result of the product of independent positive variables. For instance, in Finance Mathematics, the future price of the stock, P_N, can be modeled as the effect of independent multiple small adjustments in the initial price, P_0,

$$
\begin{aligned}
P_N &= P_0 \times (1 \pm \epsilon_1) \times (1 \pm \epsilon_2) \times \ldots \times (1 \pm \epsilon_N) \\
&= P_0 \times \prod_{i=1}^{N} (1 \pm \epsilon_i),
\end{aligned}
\tag{5.16}
$$

in which the variables ϵ_i represent the random variations in the stock value, and can have undefined probability distribution.

Therefore, applying the Neperian logarithm to both parts of Formula 5.16, one obtains

$$L_N = \ln P_N = \ln P_0 + \ln(1 \pm \epsilon_1) + \ln(1 \pm \epsilon_2) + \ldots + \ln(1 \pm \epsilon_N), \quad (5.17)$$

or,

$$L_N = \ln P_0 + \sum_{i=1}^{N} \ln(1 \pm \epsilon_i), \quad (5.18)$$

in which the second term on the right side of the equality is the sum of independent random variables.

By the Central Limit Theorem, the probability distribution of the sum, that is, the probability distribution of the random variable L_N, given that the small increments, ϵ, are independent, can be approximated by the Normal distribution. Therefore, the stock price, P_N, given by

$$P_N = e^{L_N} \quad (5.19)$$

can be considered a random variable that has a Lognormal probability distribution.

6

Microeconomics

" Anyone who believes that exponential growth can go on forever in a finite world is either a madman or an economist."
Kenneth Boulding

6.1 Introduction

This chapter presents the main economic definitions that are useful to the understanding of Microeconomics, from a theoretical point of view. The economics glossary available in the appendixes contains a good part of the terminology associated with the area.

6.2 The Concept of Microeconomics

Microeconomics is the theoretical process elaborated to determine the general balance conditions of the economy starting from the behavior of the economic agents, individuals, groups of individuals or organisms that constitute, from the economic movement point of view, the centers of decision and fundamental actions, such as producers and consumers (Chacholiades, 1986).

It deals with the way the individual entities that compose the economy, private consumers, commercial companies, laborers, huge landowners, goods producers or private services act reciprocally.

Microeconomics analyzes the market and the mechanisms responsible for the formation of relative prices, how the company and the consumer interact and arbitrate the price and the quantity of a product or service (Blomqvist et al., 1987).

It studies the interactions that occur in markets, in function of the existing information and the state regulation, to explain usual market practices, such as monopoly, monopsony, oligopoly, oligopsony, perfect competition and monopolistic competition.

The operation of supply and demand in the price formation is the basic objective of the theory, therefore, Microeconomics is divided into three fundamental study areas, the consumer theory, the company theory and the production theory.

6.2.1 The Consumer and the Demand Curve

The consumer theory evaluates the client's preferences, analyzing his behavior, his choices, the restrictions in relation to values and the market demand, and produces the economy demand curve.

The Figure 6.1 illustrates the demand curve of a certain product or service, in which the Q represents the amount put into the market and P indicates the price. The curve serves the purpose of evaluating at what price there will be demand for a certain product.

For instance, if the product is put into market at an elevated price P_A, it will possibly have a low demand Q_A, as shown. If the price falls in function of the appearance of competitors in the market, the price could fall to P_C and the demand could increase, reaching Q_C.

6.2.2 The Company and the Supply Curve

The company theory analyzes the structure and the combination of capital and work in companies, to produce products according to the market demand

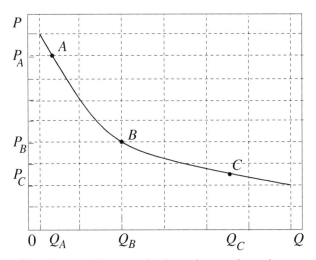

Figure 6.1   The graph illustrates the demand curve of a product or service.

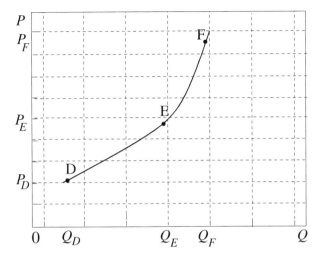

Figure 6.2 The graph illustrates the supply curve of a product or service.

and the supply of consumers willing to consume them. The supply curve of products and services is produced from this analysis.

The Figure 6.2 illustrates the supply curve of a service or product. The graph shows how many units of a certain product the salespeople want to sell for a certain price.

Supposing the price is low P_D, the supply of products will be small Q_D. An increase in the price to P_F, for instance, will lead the salespeople to offer a larger amount Q_F, in this case.

6.2.3 Production and the Cost Curve

Lastly, the production theory studies the transformation process of the production factors, those associated with the land (arable areas, forests, mines, natural resources), to labor (human being), and to the capital (machines, equipment, computers, installations), in specific products for sale in the market.

It determines the cost curves, which are used to calculate the adequate supply volume, from the relations between the variations of the production factors, to determine its effects on the final product.

Figure 6.3 illustrates how price formation works in the market. The market prices oscillate in function of the demand and supply curves, creating market surpluses, as shown in the region above the two curves, or lack of products in the market, as in the region below the respective curves.

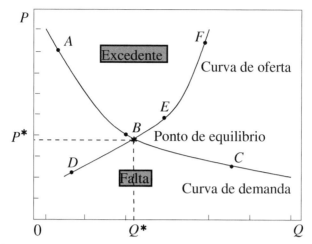

Figure 6.3   The graph illustrates the functioning of price formation.

Eventually, the market can reach an adequate price, the market break-even point for this product, given by P^*, corresponding to an amount offered Q^*.

The amount of a product that the buyers desire to acquire depends on the price established by the market, and the demand graph indicates the empirical relation, obtained from measurements or statistical surveys done in the market, between the price and the quantity.

However, that quantity also depends on other factors, such as population income. The demand curves show how the required amount is affected solely by the established price.

When the demand curve is traced, the other factors that might affect the amount demanded are kept invariant. However, eventually, the salaries can increase and this can affect the demand if prices are kept at the same level.

Figure 6.4 illustrates what occurs to the demand curve when income increases and people start to buy more typical, normal, or usual products. The demand curve moves to the right, from D_1 to D_2, with the income increase.

This way, point A_1, of the demand curve D_1, moves to point A_2, of the demand curve D_2, with an increase in the quantity, from Q_1 to Q_2, without altering the original price, given by P_A. Therefore, if the current prices are maintained an increase in the demand for the product occurs.

Figure 6.5 illustrates the change in the supply curve when a problem in the production of a certain item occurs, which is now missing from the market. This causes the supply curve to move to the left, increasing prices.

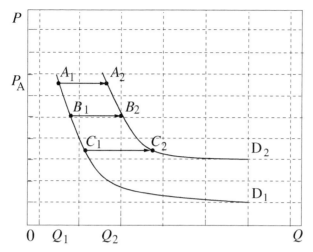

Figure 6.4 The graph illustrates the change in the demand curve for a product or service.

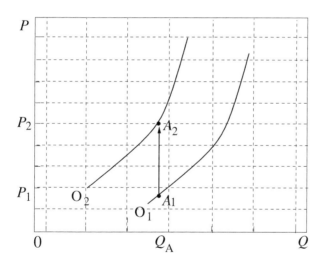

Figure 6.5 The graph illustrates the change in the supply curve of a product or service.

For instance, for the same quantity produced Q_A, with the curve moving, the point moves from position A_1, in the supply curve O_1, to position A_2, in the supply curve O_2, causing the price to increase from P_1 to P_2, as shown.

6.3 Costs Associated to Managing Companies

Companies incur several costs to maintain function. The capital cost of a company can be defined as the return rate that the company must obtain in their investments to maintain their market value unaltered.

The production cost represents the sum of the expenses referent to the raw materials, labor employed, and other expenses arising from the elaboration of a product. A company also has financial costs, the sum of expenditures referent to interest rates, corrections, and fees charged on loans and financing.

Fixed cost is the type that does not vary according to the production volume, and in general, is contractual, as in the case of expenses with rent. While the variable cost is in function of the company sales volume. Production costs are examples of this type of cost.

Marginal cost is a concept used in Economics to describe the alterations caused in the total cost for a unitary change in the quantity produced, and the operational cost represents the sum of the expenses necessary to maintain the productive serves of a private or state-owned company.

6.4 Definitions Associated to the Capital of a Company

A private company is that which belongs to individuals or groups, which aims to produce goods and services to obtain profit. A subsidiary is a company controlled mostly by another company.

A company is characterized by a certain production capacity or volume of goods and services that the company can produce during a pre-determined work period. The revenue of a company is the total value accrued from the sales of products or services.

This usually depends on the installed capacity, which is the production potential of a certain sector in the economy. To say the industry is working at 70% of its capacity is the same as saying that 30% of its production capacity is idle, the difference between the effective production volume and what would be possible to produce with the installed capacity.

Capital is the sum of all the resources, goods and values, mobilized for the constitution of a company, and can be specialized in:

- A company functioning in the anonymous society regime, and that has its shares in the hands of few shareholders, without negotiating stocks in the stock market is a closed capital company.
- Working capital is the part of the capital that is used to finance the company circulating assets, which guarantees a safety margin in the

financing of the operational activity. It refers to the capital, owned or third-party, utilized by the company to finance its production, for instance, the money used to pay suppliers. The third-party resources are usually gathered with the help of commercial banks or by means of operations such as duplicate discount.

- Working capital is defined as the difference between the current assets and the current liabilities of the company.
- Venture capital is the parcel of the capital of a company that is invested in activities or instruments in which there is the possibility of loss or gains superior to those normally expected in the habitual company activities.

  Risk is a condition in which it is possible to attribute a probability distribution, using data that forms a temporal series, while uncertainty is the condition in which there is not a probability distribution that can be associated, due to the variability of the cause of data change through time, a process called non-ergodic in Economics (de Azevedo and cei ção, 2017).
- Human capital designates the set of knowledge and information accumulated by employees of the organization, as well as the investments in education for those professionals.
- The capital account of a company, in which the value of subscribed shares was integrally received, is the paid-in capital.
- Intellectual capital represents the sum of assets of a company, which represents the knowledge total and can guarantee a competitive advantage.

6.5 Company Accounting

Accounting is a science that measures the economic reality, with the objective of studying the patrimony of economic agents, their phenomena and variations, in the quantitative and qualitative aspects, to record the facts and acts of economic and financial nature that affect the functioning of the business entities, including their consequence in the financial dynamic. The study may include:

Economic Analysis – Applying to the economic reality of the scientific method of decomposing a problem into a set of elements of easier comprehension.

Financial Analysis – Methodology based on the analysis of financial statements, such as balance sheet, income statement, and demonstration of origins and resources, of a company, with the objective of determining its current financial position and then project its possible future performance.

Marginal Analysis – Comparison between the costs incurred with the benefits obtained from some financial strategies, so that the company can better analyze its strategy in an attempt to maximize its profitability.

Balance – Financial demonstrative that lists all the assets and liabilities of a company in a certain period.

Accountability evaluates the assets of companies, which can be of several natures. Goods and services are products of economic activity or constitutive elements of production. Capital goods are those utilized in the production of other goods, such as machines and equipment.

Consumer goods are those used directly by the public in general. Durable consumer goods provide service for a long period, such as washing machines, refrigerators, and automobiles. The remainder is capital goods.

Production goods denominate, in addition to capital goods, the intermediary goods, and raw materials. Lastly, intermediary goods are manufactured goods or processed raw materials employed in the production of other goods.

Some definitions usually found in the study of Accounting are:

Capital Increase – Terminology used to reflect changes in the capital structure of a company, by means of incorporating new resources, or reserves, to its capital.

Productivity Increase – Elevation in the quantity of goods or services produced in the same area during a certain period.

Self-Funding – Parcel of the investments cost with profits accumulated from a company or economic group.

Asset – Set of goods and credit that form the patrimony of an economic subject.

Current Assets – The sum of all of a company's assets that can, short-term, up to a year, be converted into liquidity, in other words, sold as a way to increase the company capital.

Permanent Assets – Composed by the sum of the tangible goods utilized in the company operational activities and that should not be converted into money, or consumed in the course of its activities, such as properties, machinery, equipment, land.

Intangible Asset – Characterizes the assets of a company, which do not have an immediate physical representation, such as patents, franchises, names and brands.

Non-Financial Assets – Fixed assets, which participate in several production cycles, and those circulating, which are consumed or transformed in a certain production or distribution cycle.

Permanent Asset – The sum of the fixed assets of a company, such as property and machinery, with long-term investments like participation in related companies.

Public Asset – Resources in various forms, belonging to the government.

Profitable Asset – For financial institutions, it reflects the sum of all the assets that generate financial feedback to the institution. The total feedback of these assets is included in the gross revenue for the financial intermediation of the institution.

Balance Sheet – Financial demonstration that details and quantifies the assets, liabilities and patrimony of a company.

Trial Balance Sheet – Partial balance of the economic situation and patrimony of a company, which refers to a specific period in the social exercise of the company.

6.6 Business Companies

From the legal standpoint, society is a legal entity, instituted by a contract, which brings several people together which obligates themselves to employ common values, goods or labor, with a profitable purpose. Societies can be characterized as

Anonymous Society – Company formed by at least seven partners, where each one's capital corresponds to the proportion of the votes held and the liability of each partner, in the case of bankruptcy or damages caused by the anonymous society to a third party.

Limited Society – Company constituted by limited liability shares, which define the maximum amount of each partner's liability.

The capital structure of a company represents the nature of the main shareholders, the form of control, as well as the degree of concentration of equity ownership. In the case of an anonymous society, shares are launched, in other words, the subscription of debentures or other bonds by financial institutions committed to advance the capital, or acquire the part of bonds that were not absorbed by the market, or, at least, make efforts to minimize the eventual problems in selling bonds.

The debenture is a bond that represents a loan to an Anonymous Society (A.S.), which earns interest and monetary correction, being guaranteed by the company assets and with a preference for redemption over almost all the other debts.

A share is a document of property of a fraction of the capital of a society. The capital of a society can be divided into equal fractions, allowing each partner to subscribe shares according to the capital availability. The shares can be defined as:

Book-entry share – Bond circulating in the capital markets without the emission of certificates or cautions, carried by a bank, which acts as a company stock depositary and processes the payments and transfers by means of emitting bank statements.

Endorsable share – Share that can be transferred through an endorsement on the back of the caution.

Registered Share – Share with a nominal certification of its owner. The certification, however, does not characterize possession, which is only defined after the launch in the Registered Shares Ledger Book of the emitting company.

Ordinary Share – Share which grants its owner the right to vote in the society assembly.

Preferred Share – Share that gives its owner priority when receiving dividends, or in the case of dissolution of the company, the reimbursement of the capital. Does not grant the right to vote in the society assemblies

To build an anonymous society, a share fund is created, in which the resources collected with the shareholders are invested in a diversified stock portfolio, with the result of the revenue of this portfolio getting distributed among the shareholders, proportionally to the number of shares, after deducting taxes and fees charged by the portfolio manager.

Shareholding control defines the decision power of one or several shareholders over a certain company; this control is guaranteed by the possession of the largest number of shares with a right to vote. Usually, there is a controlling shareholder, the individual, company, or group of people, that by means of their partner rights, can control the votes and deliberations during the company's general assembly, in addition to naming most of the company's administrators. However, there are several types of shareholders:

Effective Shareholder – Final owner of the shares of a company, which may or may not be represented by a nominal shareholder.

Minority Shareholder – The one that possesses a number of shares with the right to vote inferior to half, which may or may not hold control over the society, depending on the voting power of the rest of the shareholders.

Majority Shareholder – The one that has, at least, over half the shares with a right to vote of a company, and holds its control.

Nominal Shareholder – A company that owns shares, but only covers up the effective shareholder, that does not wish to be identified.

Fund Provider Shareholder – Small shareholder whose application strategy is only the revenue and the stock appreciation.

The shareholder general assembly is the top decision institution of an anonymous society, which decides on mergers, incorporations, distribution of dividends to shareholders, which represents the profit parcel of a company distributed to shareholders as a form of remuneration, after a proposal by the directory

6.7 Operations in the Stock Market

Financial Mathematics is a branch of Mathematics that studies the equivalence of capital through time. Its knowledge is indispensable to comprehend and operate in the financial and capital markets, in addition to acting in financial administration.

A share represents the smallest negotiable parcel of the capital of a company. When a shareholder buys stock from a company, he acquires the rights and duties of a partner. The environment in which the transactions, the acts by means of which an economic unit manifests its participation in economic life, with stocks, is the stock market.

The stock market is a nonprofit civil association with the objective of maintaining a local or electronic negotiation system adequate to the accomplishment of buying and selling of bonds and mobile values. The Stock Markets aims to also preserve elevated ethical standards of negotiation and divulge the operations with haste, amplitude, and detail.

The stock market is a centralized market for the transaction of goods, especially commodities, primary products in a raw state with elevated commercial importance, both in the internal market as well as in the external market, including ores, coffee, cereal, and cotton.

Commodities are bonds corresponding to negotiation contracts with agricultural products, metals, ores, and other primary products in the stock markets. The deals refer to the delivery in future of goods but do not necessarily mean that there is a physical exchange of products in the stock markets.

Furthermore, there is the market of goods and futures, an institution in which goods are negotiated, especially the most important ones in the internal and international market, whose stocks can be existing or projected.

Capital gain is the difference between the revenue received from the sale of a certain asset, such as shares or property, and the cost of acquiring the asset.

Some usual market operations are:

Risk Analysis – Continuous and systematic analysis of the adverse effects, or risk, that may hit a certain company in a competitive market.

Application – Operation in which the buyer, an investor or company, acquires a part or entirety of the shares of a company.

Acquisition – The purchase of a company by another, in which both maintain their respective legal identities, as opposed to what happens in a merger or incorporation. In the case of anonymous societies, it can mean only the acquisition of a batch of shares sufficiently large to allow the control of a company.

Leverage – Improvement effect caused by the indebtedness in profitability of the net patrimony of a company. Represents the relation between the indebtedness of a company and its equity, in other words, its net patrimony.

Fictitious Accumulation – Appreciation of property titles, shares, or other financial assets, independently from the assets represented, with no relation to the effective production of wealth. Typical of a speculative process, especially in the stock markets or from a privileged association with the public fund, via stock market or governmental institutions.

The net gain, in the market, equates to the difference between the sale value of a certain financial asset and its acquisition cost. In the future market, the result is the sum of the daily adjustments that occurred each month.

The New York Stock Exchange (NYSE), was created on May 17, 1792, by twenty-four brokers that signed the Buttonwood Agreement. It is established in the Financial District of Lower Manhattan, in New York City, and it is the World's largest stock exchange by market capitalization, hosting 82% of the Standard & Poors 500, as well as 70 of the biggest corporations in the World.

The main index of the São Paulo Stock Market, which expresses the daily average variation of the negotiations, Ibovespa, was implemented in 1968, and is currently formed by a theoretical portfolio of about 70 stocks, which are chosen by stock's participation in the market and by liquidity. The São Paulo Stock Market evolved from the Free Stock Market, founded on August 23, 1890, by Emílio Rangel Pestana.

6.8 Financial Institutions

A financial application characterizes the operation by which an individual or entity transfers, temporarily, the exercise of a certain purchasing power to a financial institution in exchange for the obtainment of income on the invested capital. A financial agent is a financial institution that can represent, as a guarantor, financier or endorser, a public entity.

The institution enabled to keep money, bonds, and values for a third party is the bank. Medici Bank was one of the first financial institutions based on international commerce, founded in Florence, Italy, in 1397.

Banks can use the deposited values for loans, executing related operations, such as charges, payments, operations in foreign currency, fixed income applications, whose income value is previously established, among others.

In addition, the bank has permission to charge a premium, an amount that the buyer pays on top of the nominal value of a bond. Typically, the commercial banks can be:

Investment Bank – Bank that offers loans and financing for implementing companies.

Retail Bank – Bank that concentrates activities and services directed to a vast market of clients.

There are banks that operate in countries, such as the World Bank, an international financial organization that gives out loans to developing countries. Together with the General Agreement on Tariffs and Trade (GATT), the World Bank regulates the international financial system. It was created in 1944, and originated the World Trade Organization (WTO).

The Inter-American Development Bank is an international institution, with headquarters in Washington, USA, to help financially with development of the infrastructure of emerging countries.

In Brazil, the National Bank for Economic and Social Development (BNDES) is an entity that belongs to the Brazilian government, responsible for the execution of its long-term credit policy.

6.9 Competition or Not

A cartel is a type of deal done between competing companies, with the objective of limiting or suppressing, the risks of competition, in addition to increasing the prices of products. It aims to increase profits, at the expense of the consumers, by fixing prices or production shares, dividing clients and markets, or by the coordinated action between the participants. The formation of a cartel is not always evident (Mirow, 1978).

Monopolistic competition occurs when two or more companies that produce similar goods or services, however, not interchangeable by each other, keep a certain degree of control over the price of those products.

Capital concentration is the increase in the size of a company or capital block due to the process of accumulation. The formation of conglomerates is not uncommon, in which companies or economic groups act in several sectors of the economy, without necessarily following technical, productive, or commercial complementarity criteria. Typically, it is followed by income concentration, which is the process by which income from profit, salary, or other revenues, converges to one region, company, or restricted group of people.

Monopoly is the situation in a market in which there is no supply competition. Duopoly occurs in the particular case of an oligopoly in which only two sellers of a certain good or service exist.

Oligopoly is structured in the market when a small group of companies controls an expressive parcel of the supply of products and services.

The situation in a market where the competition is imperfect on the demand side, due to the presence of a very limited number of buyers originates an oligopsony.

7

Macroeconomics

"The economy signifies the power to repel the superfluous in the present, with the objective of securing a future good and this aspect represents the dominance of reason over the animal instinct."
Thomas Atkinson

7.1 Introduction

This chapter presents the main economic definitions and concepts that are helpful to understand the subject of Macroeconomics. The glossary on Economics that is available in the appendixes contains the terminology associated with the area.

7.2 Concept of Macroeconomics

Macroeconomics is the branch of Economics that studies, on a global scale, by statistical and mathematical means, the economic phenomena and their distribution in a structure or sector, verifying the relations between elements, such as the national income, the price levels, the interest rates, the savings and investment levels, the trade balance, exchange variation, and level of unemployment.

The first complete economic analysis was done in 1758, by François Quesnay (1694–1774), French doctor and economist, which stood out as the main exponent of the physiocrat school of thought.

In general, mathematical models are elaborated, formal representations, typically as equation systems, logic systems, or algorithms, which form coherent sets of relations between the economic phenomena (Koutsoyannis, 1977).

Therefore, macroeconomics is the study of the economic sciences under the broad perspective, explaining economic processes, which take into account aggregates and aspects such as inflation, interest rates, and exchange variation.

Represents the analysis that intends to guarantee the maintenance of the complete use of the available resources in the economic systems. Macroeconomics also encompasses the necessary conditions for economic development as well as their meanings, costs, and benefits. It aims to determine the causes and effects of inflation and the general elevations in price levels as a whole (Blomqvist et al., 1994).

Macroeconomics is typically non-experimental, in other words, it focuses on the basic observation of phenomena, since one cannot or should not, conduct controlled scientific experiments in a real economy.

Analogously to the Principle of Uncertainty in Quantum Mechanics, formulated in 1927, by Werner Heisenberg (1901–1976), any experimental interference to observe a macroeconomic event may induce an alteration in the process (Krane, 1999).

The circular flow of resources is an economic model, in which the exchanges are represented by cash, goods or service flows among the economic agents. The analysis of the circular flow is a way to evaluate the national accounts and forms the basis of Macroeconomics (Mankiw, 2010).

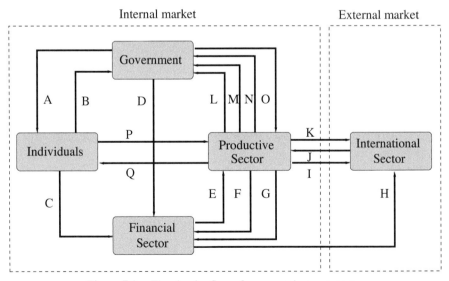

Figure 7.1   The circular flow of resources in an economy.

Figure 7.1 illustrates one of the possibilities of drawing the circular flow, for an open economy, with international interaction. It can be noted that the flows of money and exchanged goods in a closed circuit, correspond in value, but follow opposite directions. The dotted boxes involve the respective national and international markets.

The international economy deals with the flow of goods, services, and capital through the borders between the countries. The commerce of goods refers to imports or exports. The service transactions involve activities such as dispatch, travel, insurance, and touristic services executed by companies of one country to the residents of another. The capital flow refers to the establishment of plants in foreign countries, or the acquisition of bank accounts in a country by residents of another (Kreinin, 1987).

The symbols shown in the figure correspond to the following meanings:

A – Transfers done by the Government, including salaries, goods and non-mercantile services.

B – Direct taxes charged by the Government, on income and property.

C – Families' private savings.

D – Applications done by the government.

E – Gross investments.

F – Business savings.

G – Depreciation.

H – External savings.

I – Revenue sent overseas.

J – Exports.

K – Imports.

L – Direct taxes and tariffs.

M – Indirect taxes and tariffs.

N – Governmental revenue.

O – Governmental expenses and subsidies.

P – Personal revenue flow.

Q – Income, which includes salaries, rent, interest rates and dividends.

Richard Cantillon (c. decade 1680–c.1734), the Irish economist, initially presented the idea of circular flow in his book *Essai sur la Nature du Commerce en Général*, which can be translated to Essay on the Nature of Commerce in General, considered the initial mark of Economic Politics (Wikipédia, 2020b).

Cantillon became a successful banker and merchant, because of the political and business connections he formed. He helped found the Mississippi Company, with John Law, which made him rich by speculating stocks.

However, because of the losses caused to investors in this process, he was persecuted with lawsuits, criminal accusations, and assassination plans, until his death, in 1734, which some authors say was criminal (Gleeson, 1999).

Eventually, Quesnay developed the important concept of circular flow with his *Tableau Économique* (Economic Framework), shown in Figure 7.3. The idea was later improved by Karl Marx, in "The Capital, a Critique on Economic Politics," and by John Maynard Keynes, in the book "General Theory of Employment, Interest and Money."

Richard Stone (1913–1991), the British economist and Economic Sciences Nobel Prize recipient improved the concept for the United Nations

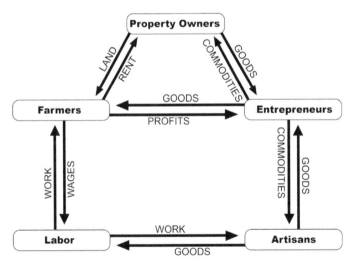

Figure 7.2 Model of circular flow initially proposed by Richard Cantillon. Image adapted from Wikimedia Commons, a collection of free content from the Wikimedia Foundation.

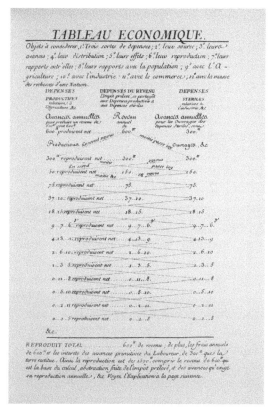

Figure 7.3 Economic framework of Richard Cantillon, which originated circular flow. Image adapted from Wikimedia Commons, a collection of free content from the Wikimedia Foundation.

(UN). In his acceptance speech, in 1984, Stone mentioned François Quesnay, as well as his *Tableau Économique*, indicating that was one of the first works in Economics to examine several sectors and their respective interconnections on a global scale.

Quesnay, to whom the expression *laissez-faire* is attributed, defended the economic freedom and created the supply and demand paradigm, in other words, the higher the demand for a product, the higher its price is. He advocated that with freedom, production and consumption is limited to what is necessary, to possibilitate stability in prices and economic equilibrium.

However, according to Claude-Camille François, Comte d'Albon, in his work *Eloge Historique de M. Quesnay* (1775), Vincent de Gournay (1712–1759) would have been the author of the famous quote "*Laissez-faire, laissez-passer, le monde va de lui-même*," which means "Let it be done, let it go, the world goes by itself. " On the other side, it is attributed to the Marquis d'Argenson the primacy of having for the first time, associated the expression to liberal doctrine.

Microeconomics is also known as the Theory of Prices, because it aims to analyze the formation of prices in the market, in other words, how companies and consumers interact and decide the price and quantity of a certain good or service in specific markets. Thus, Microeconomics deals with:

- The fundamental units of Economics, such as consumer and company, considered in an isolated or aggregate manner.
- The behavior of the consumer, motivated by the search for satisfaction, limited by an eventual budgetary restriction, among other stimuli.
- The behavior of the company, that looks to maximize profit, among other incentives, conditioned to its cost structure and oriented by the performance of its competition and by legal norms.
- The mechanisms that rule the workings of the market, especially the so-called laws of supply and demand.
- The market role and its imperfections in dealing with the efficient use of the Economy limited resources, and possible problems in the production of goods and services aimed at consumer satisfaction.
- The remunerations paid to agents that participate in the productive process and how the social income is shared.
- The definition of prices that remunerate the resources received by the units that generate the final goods and services.
- The definition of private costs and benefits, with the objective of the greater good of society.

7.3 Operations in the Governmental Scope

The governments have a responsibility for the maintenance of stable exchange, over the accounting adjustment of their accounts, over the emission of money, over the commercial balance, among other obligations, which follow:

Exchange Adjustment – Modifying prices of the national currency in relation to foreign currency. Also known as exchange correction.

Fiscal Adjustment – Set of procedures of the Federal Government with the objective of spending less than is earned.

Monetary Aggregate – Set of homogeneous elements that compose the currency supply in a country.

Exchange Anchor – Set of exchange policy measures with the objective of maintaining a fixed exchange rate, indicating that the country possesses reserves to defend the currency against speculative movements.

Fiscal Anchor – Set of measures to maintain a stable economy by means of public expenditure contention.

Monetary Anchor – Set of measures with the objective of keeping a stable economy by means of determining a limit to currency supply.

Trade Balance – Registers all the exports and imports done by the companies in a country.

Capital Balance – Accounting of all the movements relative to actives and passives, that occurred in a country and the rest of the world, in a certain period.

Payment Balance – Registers the result of all transactions, such as goods, services, transfers, and capital flows, between a country and the rest of the world.

Balance of Current Transactions – Trade balance, exports minus imports, and services such as payment of interest rates on foreign debt.

7.4 About Currency

Monetary standard is a value, or matter, conventionally adopted as a basis for the monetary system of one or several countries, and in relation to which are defined other types of currency, the monetary units. Fiduciary currency is a standard that is not backed by gold and has no intrinsic value; the government determines its price. The basis is the sum of the currency emitted and the bank reserves held by a country (de Carvalho, 2012).

Monetarism is the theoretical current that attributes to currency a determinant role in economic fluctuations, whose adepts defend the quantitative theory of currency and its implications. The gold standard was the monetary system that linked the value of currency from each country to gold, it was suspended in 1931. The increase in the use of paper money or checks is known as monetization.

7.4.1 Cryptocurrency

Cryptocurrency is a type of virtual coin that uses encryption to guarantee safer financial transactions on the Internet. There are several types of crypto coins, Bitcoin, also known as BTC, is the best known, a digital encrypted coin which allows financial transactions free of institutions, but monitored by users of the network and encoded in a (blockchain) database.

Bitcoin does not derive from physical currency; it is not backed financially and is not recognized by the Securities and Exchange Commission (CVM) in Brazil as an asset. Despite being introduced in 2009, the *de facto* asset Bitcoin only aroused interest in the media in 2012. Strictly speaking, the encrypted financial assets, or crypto assets, should not be classified as currency, since they do not meet the requirements to receive this denomination, considering the current monetary theory.

Currency is typically defined by three fundamental attributes: functioning as a means of trade, being a unit of counting, and acting as a value reserve. Bitcoin meets the first criteria, because a growing number of merchants, especially in online markets, are willing to accept it as a means of payment. However, the commercial worldwide use of Bitcoin remains reduced, indicating that few people are using it as a means of trade (Camacho and da Silva, 2018).

Blockchain is a decentralized digital database, which registers financial transactions that are stored in computers throughout the world. The database registers the sending and the receiving of the values of encrypted format digital coins, and the parties need to authorize the access among themselves.

Bitcoin originated from a process developed by Satoshi Nakamoto, who described the development of a peer-to-peer (P2P) electronic money system. The algorithm proposed by Nakamoto creates new Bitcoins and awards them to computer users that resolve specific mathematical problems. These problems get more complex and less frequent with time, because of the costs related to mining, which is the process of generating new coins.

Miners are individuals or companies that are involved in the activity to win new blocks of Bitcoins. The payments are registered in a public ledger book, and the transactions occur peer-to-peer, without a central repository or unique administrator.

This decentralized record monitors the property and the subsequent transfers of each Bitcoin, after they are extracted, or mined, by their original proprietors. The ledger is organized as a blockchain, that contains records of validated transactions to track the property of each Bitcoin.

Each transaction record contains a public key for the recipient. In a Bitcoin transaction, the proprietor validates his property using a private key and sends an encrypted instruction with his key. The system records the transaction instruction, which contains the public key of the new owner, which is the recipient in a new block. This system o decentralized authentication also serves for dealing with falsifications and double accounting problems.

The P2P network attributes parity of rights to all the members. The decentralization is maintained by combining Proof of Work (PoW) with other cryptographic techniques. PoW is a hash mathematical function, an algorithm that maps variable length data into fixed length data, that has a large number of possible results for each input.

The hash in each block is connected to the next block. The hash function can be represented in the following manner

$$h(n) = f[h(n-1), \phi, K], \tag{7.1}$$

in which $h(n)$ represents the hash of the current block, $h(n-1)$ is the hash of the previous block, ϕ is the difficulty level and K is the random specific key for the current block.

This indicates that each subsequent block is connected to the previous one, if there is a dishonest miner that decides to generate an invalid block, the other members of the network will not confirm the operation, because the hash of the previous block will be validated.

Even if a user tries to change the hash of the previous block, he would have to do it for the block before it, and successively until the very first block created by Nakamoto. This would require an enormous amount of computational work, which would exceed the current capacity of the network.

So, to protect the integrity of the ledger book, the system protects each block with a unique hash. The hash is generated based on the system information and the owner's key and needs to meet the criteria given by the hashrate defined by the system, 136 quintillion hashes per second (136 EH/s), in 2020. New blocks which document the recent transactions are confirmed and add to the blockchain only when a valid hash is found.

Simply put the process of creation of BTC works in the following manner. The data of a transaction involving BTC are transmitted to all those participating in the (P2P) network and, so that the transaction can be made possible, it needs to be processed, and in other words, the cryptographic problem must be solved.

The miner receives 12,5 BTCs for each transaction block discovered, a payment for having loaned computational power to enable the transactions in Bitcoin. The miners can also be rewarded with a fraction of the transactions done, which can be offered optionally by BTC users.

The rates are offered so that the transactions are prioritized by the miners in the formation of the candidate blocks, increasing the processing rate of the transaction (Camacho and da Silva, 2018).

7.4.2 Financial Operations with Currency

The external accounts form the accounting system that registers the economic transactions of the residents of a country abroad. An exchange crisis is an adverse economic situation of a country, which can lead to the rapid loss of its exchange reserves. Some definitions and typical operations are done with a currency, such as the following:

Exchange – Operation in which the national currency is exchanged for foreign currency or vice-versa.

Commercial exchange – Dollar quotation, used for closing contracts on exports and imports. The commercial exchange also records the foreign company loan operations, direct foreign investments, foreign investment entries and exits in fixed income and in the stock market.

Fixed exchange – Exchange system in which the central bank of a country establishes a fixed price for the parity between the local currency and the dollar, or euro, for instance.

Fluctuating exchange – It is the market in which multinational companies ship their profit abroad. The fluctuating exchange also includes loan operations and jewelry and precious gems business.

Parallel exchange – It is the market that exists when the country does not have a completely free exchange policy.

7.5 Public Accounts

The budget represents the set of the government's daily expenses. The necessary expenses for the ministries, departments, and public administration agencies to function. This type of expense appears as other current expenses and does not include expenses with salaries.

The services account records the income and expenses with international trips, debt expenses, freightage and insurance, and shipments of profits and dividends from companies.

The tourism account shows the citizens' expenses on international trips and the income received by the country when foreigners visit the country.

A financial crisis is a situation that reflects an imbalance between the income and expenses of an entity. The set of goals from the accord with the International Monetary Fund (IMF), by which a country is evaluated, is denominated performance criteria. The non-compliance of these foals means the interruption in the transfers of the loan installments and demands a new renegotiation.

7.6 Public Debt

The State is usually the biggest debtor of the financial system. The debt is typically generates a deficit in the governmental accounts, producing a negative result when the expenses are subtracted from the income. Fiscal imbalance is the situation in which the government's expenses surpass the income collected by means of tributes and other permanent income sources.

It is possible to identify two main sources for the supply of capital in an economy, individual savings and company savings, called S, and the influx of capital from foreign investors, which is equal to the commercial deficit $I - E$, in other words, imports minus exports.

There are also two primary sources for the demand of financial capital, the private investment sector, P, and the collection of loans by the government, which occurs when governmental expenses, G, surpasses the value of the fees and taxes collected, T.

Therefore, it is possible to express the identity between the national savings and the investment by the following equation (Greenlaw and Shapiro, 2018),

$$S - (I - E) = P - (G - T), \tag{7.2}$$

that translates the idea that the supply of financial capital equals its demand, in which, again, S is private savings, I represents the imports and E the exports, P is the investment, G is the governmental expenses and T represents the fees.

The commercial deficit reflects the difference between what the country collected from exports and what it spent on imports. When the result is negative, the imports are greater than the exports, characterizing a commercial deficit. If the result is positive, a commercial surplus occurs.

Checking account deficit is the of commercial transactions of a country with the world, including exports and imports, plus services, the so-called unilateral transfers, insurance expenses, maritime freightage, and the other expenses with foreign commerce.

Because of the higher value of public expenses in relation to the total tributes collected by the government, a fiscal deficit occurs. When including the expenses with public debt interest rate payments, in addition to the income and expenses, the nominal deficit is produced.

The operational deficit is the resultant of the higher value of current expenses, investments, and interest rates in relation to the current income. The primary deficit is the result of negative public accounts, including the National Treasure, pensions, and Central Bank, disregarding the public debt interest rates.

The negative result of public accounts, resulting from the excess of public expenses in relation to the resources collected by the government produces the public deficit. In addition, there is a pension deficit, the difference between what the government collects from the public employees and what it pays as benefits to the active and inactive public servers.

The external debt represents the total debt to the government's foreign creditors, including States, Cities and state-owned companies. The internal debt is taken on between the government and individuals and companies residing in the country.

Liquid debt is the sum of all the financial obligations, like loans, debentures, fixed-income bonds, from a company, short term or long term, with the company availabilities deduced, in other words, the sum of the instruments that can be considered paper currency.

The volume of bonds that the government emitted and sold to the market characterizes the securities debt, and the public debt is everything that the government spends with loans and bond emissions. Lastly, securitized debt is the responsibility bond of the National Treasure emitted because of the assumption and renegotiation of the debt the government took on by force of law.

7.7 Taxation

Taxation is the system by which the Federal Government, the States or the County collect taxes. This can eventually lead to a fiscal gain, which is the advantage obtained by the government because of the increase in tributes or the reduction of expenses.

Tribute is the income instituted by the Government, the States, the Federal District and Counties; it includes taxes, fees, and improvement contributions, in the terms of the Constitution and the current laws on financial matters.

The tribute charged on income and salaries is called income tax, with different values for each type of income, and also in relation to individuals and companies.

Arthur Laffer (1940-), the American economist, defended in 1974, that the increase in taxes could cause a reduction in revenue. The relation between the collected value of taxes by a government and the possible taxation reasons was presented as a curve, which took his name.

The dismissal of tribute payment granted legally to some products, services, or activities, known as a tributary exemption, is granted to certain products, for a certain time, as a way to stimulate the Economy. However, the economic zone where the fiscal and monetary regulation of banking activities is light or nonexistent is known as a tax haven.

7.8 Remuneration of Capital

Interest rates are the cost of money in the market, regulated by the Central Bank. Interest rates are calculated by dividing the fees paid per year by the original amount of the loan. The basic interest rates are the annual rates determined by a bank, which serve as a reference for the calculation of the different conditions offered by this bank.

Reference rate, known as RR, is the interest rate used as a reference, calculated by the Central Bank of Brazil (BC), used as an indicator for contracts or financial assets.

In addition to the usual rates, late fees are also applied, known as late payment fees. The term defines the interest rates charged by credit card administrators in the case of late payment.

Real interests, that affect a loan or financing, do not include the monetary correction of the loan amount. The nominal interest corresponds to interest rates, which include the monetary correction on the loan value.

The future interest rates represent contracts negotiated in the Commodities and Futures Exchange (CFE), in which the investors bet on the trend of interest rates in the future.

Usury consists in the act of charging higher interest rates higher than the usual or permitted by law, in the case of a loan.

7.9 Profit and Capital Investment

The capacity of transforming a certain asset or investment in money is called liquidity, and it also indicates the volume of money circulating in the market. If the liquidity is high, for instance, there is a lot of money circulating through financial institutions. In the financial market, the term determines the capacity that a bond has of being converted into money.

General liquidity is a financial analysis indicator, utilized to measure the liquidity of a company, that shows how much a company will receive in relation to what it owes, encompassing the long-term assets and liabilities.

Current liquidity determines that amount a company will receive short-term, in relation to each monetary unit that it will pay in the same period.

On the other side, dry liquidity reflects the capacity of a company in accomplishing its short-term obligations, with the difference that stock is excluded from the circulating assets of the company.

Profit represents the gains received from an operation in a certain period of economic activity. It can be gross when considering the variable expenses, such as taxes, or liquid when it discounts the taxes. Operational profit is the profit due to the productive operations of the company.

7.10 Current Market Economy

The Market economy is a system in which the economy is controlled by private initiative economic agents, which act freely, with a minimum intervention by the State. In the market, the economy is regulated by supply, which is the offering of goods and services to the market, usually for a certain price and time, and by demand, defined as the amount of goods and services that can be acquired by a certain price, in a given market, during a certain amount of time.

The market, then, defines the price of goods or services. This price is the monetary value expressed numerically associated with a good, service, or patrimony. Independent of its objective use value and its subjective satisfaction value, the price of a good or service only exists when it is situated in a trade relation.

In the open market, the monetary authorities, such as the Central Bank, function with public assets, to regulate and control the payment means, at the same time as they finance the internal federal debt.

In the spot market, the physical liquidation, the delivery of the assets by the seller is processed on the second business day after the floor trading, and the financial liquidation, the payment of the assets by the buyer, happens on the third business day after the negotiation, only through the effective physical liquidation.

The foreign exchange market integrates a set of institutions that accomplish the conversion of national currency into foreign currency, and vice-versa, in operations related to the foreign commerce of goods and services.

The capital market constitutes a set of institutions that provide long-term capital. This includes the Stock Market, banks and insurance companies. The stock market forms a segment of the capital market that comprises the primary placement of new stocks emitted by companies and secondary negotiation, on stock markets and over-the-counter markets, of the stocks already in circulation.

The credit market is a set of banking and non-banking institutions, which provide short-term capital, for both consumption and company working capital, in addition to operating in the public asset market.

The futures market is where one can buy and sell to the future. The stocks can be bought for liquidation on a future predetermined date.

Call option for stocks is a type of contract that guarantees to its holder the right to buy a batch of shares at a set price, in the options market, for a determined time. The counterpart of the contract, the launcher, takes on the commitment of selling the batch at a set price, until the due date, if the contract holder wishes to exercise his right.

The monetary market is a set of institutions that accomplish short-term and very short-term credit operations for the coverage of momentary disengagements of the economic agents, including the National Treasure and the banks. It serves as the main instrument to regulate the liquidity of the economy for the Central Bank, by means of buying and selling public assets.

Part of the active economy involves goods and services banished from a certain region, for instance, the non-registered weapons commerce, illicit drugs commerce, or counterfeit products commerce. It is the so-called black market.

7.11 Economic Policies

An economic policy, also known as normative economy, represents the set of practical actions the government takes with the intention of conditioning, and conducting the economic system, to achieve certain politically established economic objectives. The Real Plan, for instance, was a stabilization plan implanted in July 1994, by the then President Itamar Franco, based on the creation of a stable currency, the real.

Exchange policy is a government economic policy that determines the value of the exchange rate and the functioning of the exchange market. The interest rate policy, also established by the government, aims to keep the interest rates at a certain level.

Fiscal policy deals with the government's income and expenses, which include the tax burden on individuals as well as on companies, in addition to defining the government expenses based on the amount of tribute collected.

The monetary policy establishes the control of money in circulation in the market, to define the interest rates. In this manner, the government controls the monetary supply, which represents the set of credits formed by the monetary and almost monetary availabilities.

The means of payment represent the resources considered immediately available to the population. It is measured by means of the money in the power of the public, plus the deposits in cash in commercial banks, including in Banco do Brasil.

Production is the creation of a good or service adequate to the satisfaction of a need. The internal production represents the set of goods and services produced by the national economy, in national territory, not matter the nationality of the producers.

The national product is the aggregate that embraces the set of products from several sectors of the national economy during a certain period, usually one year. The sum of all the goods and services produced in the country during the year in the GDP.

7.12 Income and Revenue

Income represents the sum of all the values received in a certain period. To a company the income is formed by the sales, by the part received referent to credit and by eventual revenue from financial applications.

The gross revenue is that which occurs in a certain accounting period, while the net revenue results from sales, minus the sales and production taxes, returns and discounts.

Tributary revenue is collected by the government by means of taxes, fees, and contributions. It depends on the national income, the representative aggregate of the national resource flow in goods and services, generated during a certain period. It includes salaries, liberal professional income, private profit and profit obtained by public companies, interest rates, rent, and income resulting from a lease.

Nominal income is the profits obtained from a financial application, usually tribute, without discounting eventual inflation rates. The non-taxable income is the type of profit exempt from the charge of income tax.

Per capita income is the result of the division of the total amount of taxable income, by the number of people, in an economy, an indicator used to measure the level of development in a country.

Fixed income represents the income informed to the investor during the application act, indicating how much one will profit and when the money returns, while post-fixed income is the revenue that pays monetary correction in the application period, plus interest rates, on the corrected value of the application. In this application, the investor only knows the yield at maturity of the bond.

The bond whose remuneration is not determined in advance is called variable income. The profitability of this application depends on the market conditions.

Profitability expresses the appreciation or depreciation of a certain investment in percentage terms. Average profitability is verified by the average revenue percentages obtained from a financial application during a certain period, while the net profitability is the percentage of revenue from a financial application, discounted the taxes and fees. The rentier is the individual that obtains revenue with capital applications.

7.13 International Transactions

The foreign exchange reserves are the assets in foreign currency and precious metals accumulated by a country, which function as a type of insurance for the country's obligations. They are the international reserves of the country.

There is always an element of uncertainty, called risk, which can affect the activity of an agent or the development of an economic operation. The exchange risk represents the risk possibility for the investor, caused by the foreign exchange policy instability of a country.

The current transactions balance is the result of all the operations of the country with foreign countries, including the income and expenses from the trade balance (exports and imports), from the services account (interest rates, international trips, transportation, insurance, profits, and dividends, miscellaneous services) and from unilateral transfers. The difference between imports and exports is the trade balance.

8

Taxes and Income Distribution

"Economics is just the art of reaching poverty with the help of statistics."
Roberto Campos

8.1 Introduction

There are constant complaints about the collection of taxes by governments and the money application, all over the world, but it is important to know that the business sector has one of the largest tax evasion rates. Interestingly, those who most complain are the ones who evade taxation at all costs.

This chapter presents an investigation on how companies use the economy to exploit their most fragile economic agents, today and in the future.

8.2 About Fair Taxes

A basic economic analysis indicates that, if the State took care of communications services, energy, mineral exploration, health, education, highways, water and sewage, as well as the police, justice, administration public goods, and legislation, it would be possible to substantially reduce the tax burden.

With the collection of taxes for energy, communications, mineral resources, and highways, the government could charge less taxes. This would happen, because the taxes are considered the fairest ones, since they are levied against the provision of services.

To understand the reasoning, it is important to realize that there are few ways for the government to obtain resources. The most important are (Alencar, 2016b):

1. Tax collection.
2. Currency issue.
3. Bank loans.

4. Issuance of redeemable papers.
5. Service charge.

The first method, taxation, is unpopular and, in the case of entrepreneurs and professionals, subject to tax avoidance, the technical term for tax evasion. Entrepreneurs, in general, do not like to pay taxes, although they enjoy most of the government benefits.

The second procedure, that is, issuing money, is one of the main causes of inflation, and governments generally avoid it. Only the United States can issue unsupported currency indefinitely, for the time being, because that country currently prints the world's money, even though it has abandoned the convertibility of the dollar in 1971, which led to the collapse of the Bretton Woods system and made the dollar a fiat currency, that is, non-convertible.

It is interesting to note that David Ricardo was an advocate of currency convertibility, which had also been suspended, in 1797, due to the devaluation of the notes in relation to the price of gold in England. At Ricardo's suggestion, the Bullion Committee recommended the return of convertibility, which occurred in 1821 (Ricardo, 1996).

Taking bank loans is extremely risky for most countries because the interest charged by banks is always abusive. Any government would break down, in a short time, if it only used this expedient to finance its expenditures. Notwithstanding this warning, most countries resort to bank loans to finance construction and development.

For example, interest expenses in Brazil totaled US$ 66 billion in 2019. The federal public debt, which includes government indebtedness within Brazil and abroad, increased from 9.5% in 2019 to US$ 848 billion, according to data from the Brazilian Treasury Secretariat. The interest rates take a big chunk of the country's revenue and put the future of the people at risk.

In the case of the Brazilian Federal Government, which is the largest borrower, the internal interest rate is calculated based on the rate of the Special Settlement and Custody System (Selic), which represents the basic interest rate of the economy. Selic is the main monetary policy instrument that the Brazilian Central Bank (BC) uses to control inflation. It influences all interest rates in the country, such as the interest rates of loans, financing, and financial investments.

The fourth way, that is, issuing redeemable papers, is important, but limited, because the market usually saturates and always demands higher yields to buy the papers that are auctioned by the government. For example, Redeemable Preferred Shares (RPS) are used in fundraising, especially for investments in infrastructure (Jairo Laser Proclanoy and Paulo Cesar Delaytl Motta, 1992).

The fifth measure, the introduction of service charges, does not cause controversy or strangeness in the population, since, as mentioned, the fee is charged against the provision of a certain service, such as telephony, energy or data transmission, for example. It is generally seen by the population as part of the purchase of a product.

Unfortunately, several governments of a more neo-liberal character, sold several important state-owned companies all over the World and thus waived the collection of the corresponding fees, passing these resources on to the entrepreneurs who acquired the companies.

By the end of the 1990's, entrepreneurs of the communications sector acquired state owned companies in Brazil for modest amounts, typically financed by government banks. Because of those acquisitions, the newly privatized companies began to amass a total of US\$ 46 billion in service charges per year, even with poorly rendered and excessively expensive services.

It is noteworthy that the government sold the telecommunication operators for less than it had invested only in the modernization of the same companies. If that revenue amount were appropriated by the government, it would not have cash problems in the foreseeable future.

8.3 Income Distribution in the Economy

The neoliberal wave that has plagued the world since the Margareth Thatcher era, concentrated wealth at a level never seen before. Private income today is more than six times the GDP, the sum of all goods and services that are produced in a year, by the countries of the world (Alencar, 2017a).

And this refers to absolute GDP, because the net GDP, that is obtained by subtracting the countries' public debt, is practically null, since a large part of the countries is very indebted.

Since Thatcher's time, the conservative, as well as, corrupt rulers are selling state-owned companies to greedy capitalists, who proclaim the end, or weakening, of the state, but, cleverly, get along with it. The objective of many entrepreneurs is to obtain a public concession, a sinecure, to cease investing or creating.

That is one of the reasons why a small group of only 62 capitalists has, currently, a total income equivalent to everything that has been accumulated by 3.5 billion people, that is, half of the Earth's population, during their entire life.

The excessive concentration of wealth decreases the circulation of products and money, because the purchasing power of the population is reduced. Since those capitalists mainly accumulate resources or spend on luxury objects, such as, gold and jewelry, or expensive services, such as traveling, with little added value or little usefulness, the economy becomes stagnant.

An absurd reasoning helps to better explain the situation. If all the wealth in the world were concentrated in the hands of just one person, he/she would have no one to sell to, nor anyone to buy from. And the rest of the population would not be able to buy, for not having money. The absurdly rich single person would never find someone to sell to, for the same reason. Thus, there would be no circulation of goods and services and there would be no market.

On the other hand, if the wealth were uniformly distributed throughout the world's population, everyone would have resources to buy, and also who to sell to. With that prospect, there would be a large circulation of resources in the economy, encouraging the industrial, trade, service, and agricultural sectors. Therefore, the market would be dynamic and strong and the economy would grow.

Therefore, capitalist accumulation can curb development and bankrupt the world. The socialist distribution of resources, however, with an adequate set of rules to guarantee the protection and expansion of the market, may be the salvation of the free enterprise. Of course, it is only possible with the existence of a strong state.

8.4 We Are All Bank Employees

There was a time when banks acted as financial institutions and did their businesses by providing services to customers, lending money, selling insurance, offering investment options, and accepting money for savings accounts (Alencar, 2017c).

In general, there were attendants, cashiers, managers, accountants, economists, administrators, office-boys, bank employees, everybody at the service of the customers. The customer would arrive at the bank office and define its loan or investment portfolios with the help of the bank professionals.Life used to be that simple for the customers.

Time has passed, and banks have automated their services, initially with the installation of Automatic Teller Machines (ATMs), then with the provision of automated services, and finally, with the use of services over the Internet.

Customers did not notice, at first, but in a service sector, automation does not really mean what the term proclaims. The tasks, that were once provided by the bank employees, were not forwarded to computers or robots, but they just began to be made by bank customers.

That is right, the banks gradually passed by the tasks, which were made by the attendants, cashiers, and managers, to their customers. The customers are those who have the current tasks of choosing, from a menu on their computer screens, the service to be provided, which can be the case of a bank transfer, a bill payment for a purchase, or a savings deposit.

That is, customers have become unpaid bank employees. They themselves execute the tasks that were previously done by the regular bank employees, who received salaries to do that. Banks are like restaurants of the self-service type, in which the customer assembles his own dish.

The surrealism of it all appears when the bank customer realizes that, today, everyone is, in a certain sense, employed by the banks – but, strangely, the customer continues to pay for the bank services. Therefore, with the use of computer networks and automation, soon, banks will no longer need employees.

8.5 About Paintings and Computers

When enjoying an art exhibition, tourists or buyers usually do not pay attention to the fact that a picture is basically composed of wood or other substrate, some hardware, a little glue, fabric or paper, and paint, which form its physical or material part, with low added value, usually (Alencar, 2018c).

On the other hand, the painting typically reveals images, generated from abstract concepts, from ideas produced by the painter's creative ability, which define its, immaterial or virtual, content that has the power to add a lot of value to the work of art.

In the areas of computing and communications, some decades passed before the virtual and the material were properly separated, and had their values priced, as is done in the art sector.

At the beginning of computing, the machines of the time, such as the Electrical Numerical Integrator and Calculator (ENIAC), developed at the University of Pennsylvania, basically consisted of hardware, heated by valves. They were so costly, that only governments could develop or acquire them.

Computers were, at that time, in an almost surreal way, programmed in hardware, with the exchange of wires between valves, to obtain a certain function, such as ballistic calculation, the breaking of a cryptographic code, or determining the trajectory of a rocket.

Even when computers became commercial, they were rarely sold. The big manufacturers, which were very few, such as International Business Machines (IBM), just rented their machines, a procedure known as leasing, because the cost of the hardware was colossal.

The software was minimal on that gigantic machine, known as the mainframe, which, in literal translation, means the main frame of the room. It consisted of an operational system, to control the most basic functions of the computer, such as input and output data, and perform command interpretation.

There were few application programs, such as the Common Business Oriented Language (Cobol), for business applications, commercial or corporate activities, and the Formula Translation (Fortran), for scientific and engineering work.

Over time, operating systems have become more elaborate, UNIX and Windows appeared, and new application programs have been developed. The computers became physically smaller, although more powerful, and could be purchased by ordinary users, due to the competition produced by the multiplicity of manufacturers.

Therefore, as if the art market had begun influencing the computer sector, the programs became well-finished products, and also much more expensive. The applications received an artistic treatment, and soon the form came to rival the function, in the minds of the consumers.

Some programs, deemed the sophisticated ones, have become so expensive, that they are not even sold by companies anymore, but only licensed, that is, rented for a certain period. The hardware, in comparison, is much cheaper, just like the material used to compose a painting.

8.6 The Material and the Virtual

Since immemorial times, the art world has always understood and separated well two important concepts: the material and the virtual, which represent the substrate and the creation (Alencar, 2018b).

In music, for example, a good artist can be a virtuoso, even when playing an instrument, his equipment of work, that worths only a fraction of the price of any famous song that makes up the program of a musical.

In communications, at the very beginning of telephony, equipment was fundamentally hardware. The telephone had nothing to do with programming. In fact, it did not even have a memory to store the numbers.

All the software needed, including the procedure for dialing and digit memorizing, was in the mind of the user or the operator. Some virtuous telephone operators, much in demand at the time, could even memorize all the phone numbers of a big city, such as Rio de Janeiro.

The cell phone, when it was launched in 1979, was little more than a transceiver, a combination of transmitter and receiver, with a keyboard, in addition to the electronic circuitry, the microphone, and the speaker. It had some memory to store the numbers, and had a limited processor, which ran a very basic operating system.

Over time, the cell phone has added new functions, a camera was incorporated, then came the camcorder, the calendar, the calculator, and the screen began to grow, to occupy the entire device. The processor was already much more powerful than any mainframe of a few decades before.

The number of computer programs available today for cellphones is practically countless, for any type of application the user can think of, with new applications (apps) popping up, literally, all the time.

Changes in hardware have become so frequent, that led to the creation of the concept of Software Defined Radio (SDR), that is, even the basic functions of the device, once performed on hardware, such as modulation and audio processing, are now carried out by computer programs, which run in standardized platforms.

This process, which is called virtualization, also reaches the base stations, which control connections in the access channel to the cellular system. It has also arrived at the telephone exchanges, responsible for switching and direction of calls, which, incidentally, for some time already were called Stored Program Control (SPC) exchanges.

Finally, virtualization reached the telephone network itself, formed by the junctions of optical fibers, radio links, and the set of satellites that orbit the planet Earth.

Because, as they say, life imitates art, communications engineering went on looking for the lost time. Gradually, the physical layer, the material substrate, or hardware, yields to the virtual, to the immaterial, or software.

8.7 Economic Differentials for Development

One of the principles of electricity states that current circulation in an electrical circuit occurs because of the difference in potential between any two points. The flowing current, resulting from the electronic movement, hampered by resistance, or impedance, between the points, feeds the devices and makes them work (Alencar, 2008a).

This principle is due to the work of the French Engineer and Physicist Charles Augustin de Coulomb (1736–1806), in relation to the force exerted on electrical charges, as well as the research done by the German physicist Georg Simon Ohm (1789–1854) with the electric cell, invented by the Italian Physicist Alessandro Giuseppe Antonio Anastasio Volta (1745–1827).

An information system works in a similar way. In that case, the information differential, represented by the highest value of the entropy at one of the points, causes the information to flow from one point of the system to another. This is done according to the transmission ability of the communications channel, which depends on the probability distribution of the transmitted symbols and the amount of noise in the system.

Entropy, in the context of Information Theory, was defined initially by Ralph Vinton Lyon Hartley (1888–1970), in the article "Transmission of Information", published by the Bell System Technical Journal, in July 1928, ten years before the concept was formalized by Claude Elwood Shannon (1916–2001), one of the most brilliant scientists of his time.

Similarly, in order to promote the homogeneous development of a region or country, it is necessary to create development poles, with different productive arrangements in each city or region. A model that differs from that one that has been practiced in the last decades in most countries, in which the concentration of industries in big cities has been the objective.

The economic differentials, that can be created between the coastal line and the countryside, for example, could foster the exchange of products, to promote the growth of road, rail, and air networks, the development of logistics systems, among others, to increase, uniformly, the country's infrastructure.

The concentration of companies at a certain point in the country was a requirement to establish the critical mass, necessary for the creation of research institutions, technological centers, and universities, in addition to the leisure structure associated with the capacity to produce and generate wealth. However, it also created a resource drain that ended up impoverishing other areas.

In order to achieve a country's real production capacity, it is necessary to stimulate a uniform development, by encouraging the creation of complementary companies at different points. Then, it is just a matter of letting the difference in potential, or entropy, do the rest.

8.8 Mobile Economy

Cellular mobile communications gave more freedom to users to chat on the phone while on the go. This changed habits and created a new economic niche, which flourishes with each new day. Mobility is a distinct feature of animals, but that remained unexplored in the telephony area for a long time (Alencar, 2008b).

It is interesting to note that the word telecommunication comes from Latin term *comunicare*, which means to share, confer, exchange opinions, become common, combined with the Greek prefix *tele*, which translates as distant or far. *Comunicatio* has the meaning of participation, and to communicate is to participate, to establish a flow of messages or information.

Remote electrical communication was the motivation for the development in the area, since the beginning of the telegraph with Samuel Finley Breese Morse (1791–1872), leading to the fixed telephone invention by Alexander Graham Bell (1847–1922), to the wireless telegraph invention by Father Roberto Landell de Moura (1861–1928), whose photograph appears in Figure 8.1, to the broadcasting industry, to the digital telephony, and to the Internet.

Communication has always been nomadic, but not always mobile. When, in 1947, researchers at the Bell Laboratory, which at the time was part of the American Telegraph and Telephone Company (AT&T), created the concept of cellular telephony communication, the telephone became mobile, and there was a leap in the quality of life for users, an explosion of creativity for scientists and engineers, and a gold mine for companies.

Interestingly, part of the inefficiency of the production process in several countries is a result of the low mobility of their population, especially the most needy. In poor countries, for instance, farmworkers have to wake up very early and walk, ride or drive, sometimes for hours, to arrive at work. This ends up subtracting from these citizens the possibility to study, improve, and also the right to leisure. For most companies, a worker without mobility earns little, has a low level of education, and has little to add to the production process.

Figure 8.1   Roberto Landell de Moura inventor of the telegraph and cordless telephone. Image adapted from Wikimedia Commons, a public domain collection with free content from the Wikimedia Foundation.

An alternative to providing mobility to the population of cities in the countryside would be the creation of bicycle production units, in which the products would be sold at production cost, with no profit at all, or even subsidized by the government. Workers and their children, with their own means of locomotion, would gain precious hours to spend with their families, or at school.

Initially, small workshops in the countryside would be encouraged to become bicycle assemblers. Some metallurgical companies, installed in larger cities, would produce parts for these shops. The official, or commercial, banks could finance the shops as the first step towards industrialization.

Over time, with the right incentive, bicycles manufacturers could start to produce motorized bicycles, because of the low cost of living of many small cities in the countryside, further increasing the mobility of the poorest population. The interior of the countries would build factories, spread over many places, and the economy would gain in competition, which is always healthy, but not always well received by entrepreneurs.

Economic differentials would be created, in this sector, to allow the flow of wealth. The countryside could share the results of improved economic conditions and make the mobility of large cities common in small villages. Peasants would then be able to participate in the development of the country, even from a distance.

9

Analysis of the Economic Results of Privatization

"People of the same trade seldom meet together, even for merriment and diversion, but the conversation ends in a conspiracy against the public, or in some contrivance to raise prices."
Adam Smith

9.1 Introduction

Privatization is the transfer of activities and production from the public sector to the private sector. The Margareth Thatcher government, in Great Britain, and the Fernando Henrique Cardoso government, in Brazil, are distant in time, and it is now the time to analyze the real reasons for the privatization of state-owned telecommunications companies, carried out by both liberal rulers (Alencar, 2018d).

9.2 Privatization in Great Britain

In Great Britain, a large proportion of the major industries were owned by the State, prior to 1980. Most of the companies were taken into public ownership under the postwar labour government of Clement Attlee. The principal reason for this, apart from an ideological commitment of the socialists to the common ownership of the means of production, was to use state control of those companies as a means for national economic planning and postwar reconstruction (Veljanovski, 1988).

British industry needed rationalization and modernization to recover from the devastation produced by the Second World War, and most economists believed, at the time, hat recovery was most effectively achieved by the government expenditure and nationalization, as recommended by John Maynard Keynes.

Nationalization created large enterprises which, for the most part, adopted a goal of universalization of services and, as a result, became dominated by engineers instead of businessmen. Of course, nationalization is not synonymous with state-ownership, in the sense that these enterprises were run directly by the current government.

It is not clear that the Thatcher program of privatization has been a well thought and coherent program, with a consistent set of objectives. Only in 1983, a few years after the beginning of the privatizations in Great Britain, the Financial Secretary to the Treasury stated what the objectives of the privatization program were.

Among the most important goals of the program, one can mention the following (Veljanovski, 1988):

- Reduce government involvement in the decision making of industry;
- Permit industry to raise funds from the capital market on commercial terms and without government guarantee;
- Raise revenue and reduce the Public Sector Borrowing Requirement (PSBR);
- Permit wider share ownership;
- Create an enterprise culture;
- Encourage worker share-ownership in their companies;
- Increase competition and efficiency;
- Replace ownership and financial controls with a more effective system of economic regulation designed to ensure that benefits of greater efficiency are passed onto consumers.

It is interesting to point out that the debate over British privatization has not been carried on in terms of a refined analysis of its underlying premises, or in terms of the claim that private enterprise is more efficient than public enterprise. Furthermore, privatization was seen as the enemy of competition, because the government kept the utilities intact to maximize the revenue that a sale of the assets could yield.

The real impact of privatization in Britain was not to withdraw the state from economic activity, but to change its role from a producer to the protective state, based on the dubious principle that it is not the legitimate function of the state to be involved in economic production. The major constitutional question raised by the Thatcher government's privatization program was the way it redefined the role of the state without real political debate (Veljanovski, 1988).

At that time, Thatcher's party called for labor union reforms, less government intervention in the economy, less government spending, and lower taxes. Notable among her accomplishments were (Parker, 2003):

- The promotion of the neo-liberal philosophy of government which has to a large extent replaced the social democratic orientation of government. Thatcher introduced neo-liberalism in Britain, based on the work of Friedrich von Hayek (1899–1992), an economist and philosopher from Austria;
- The restriction of trade union leaders' *de jure* powers and the breaking of those trade union leaders' *de facto* power;
- The privatization of nationalized enterprises in the areas of coal, iron and steel, gas, electricity, water supply, railways, trucking, airlines and telecommunications;
- The privatization of public housing;
- The reduction of income tax rates;
- The institution of monetarist monetary policy, with a strong emphasis on controlling inflation;
- The restriction of local government spending and the reform of local government finance;
- The closer integration of the British economy with those of the European Community.

To protect the consumer from monopoly abuse until competition developed, a new telecommunications regulator, the Office of Telecommunications (OFTEL), was created. Initially, during the planning of British Telecom's (BT) privatisation, it seemed that the government intended to regulate BT using normal competition law. But because of concerns from the Office of Fair Trading (OFT) about the expected workload and need to build up specialist telecommunications regulation expertise, the decision was taken to include in the Telecommunications Bill provision to establish OFTEL (Parker, 2003).

The BT privatisation was a success, in the sense that the Initial Public Offer (IPO) was greatly oversubscribed. The government mounted an advertising campaign to attract small investors, and the result was a success. Instead of the expected public opposition to the privatisation of BT because, in a sense, the public already owned BT as a state enterprise and therefore could object to being asked to buy shares in the company, the public backed the sale through share buying.

The outcome was a landmark in the British, and later in the Brazilian, privatisation experiments, for a number of reasons. The sale of BT established the principle that the public utilities could be sold off in spite of their size. Second, OFTEL became the regulatory model for later sector regulatory offices for gas (Ofgas), water and sewerage (Ofwat), electricity (Offer), and the railways (ORR). Finally, the sale proved that small investors could be attracted if the shares were sold at a discount.

That privatization model helped to meet the Conservative¿s objective of creating a share-owning democracy, as a stronghold against socialism. Later on, privatisations included provisions that favoured small shareholders.

In several United Kingdom public utilities prices have fallen since privatisation, reflecting gains in productive efficiency. The following example was selected to reflect the general nature of the price changes after privatization.

In the telecommunications sector, from 1984 to 1999, the average real charges fell by around 48% on average; although this price change certainly results from technology and competition, in addition to an ownership change and regulation.

Apart from some competitive tendering for contracts, such as for cleaning hospitals and schools, and isolated examples of private companies being brought in to sort out under-performing educational and health services, the British welfare state remains solidly state provided (Parker, 2003).

London Underground, by contrast, has been currently subject to restructuring in order that private companies become responsible for the infrastructure. This project was highly controversial but was promoted by the Labour party as an example of the benefits of Public-Private Partnership (PPP).

Indeed, most privatisations under Labour have taken the form of PPP's or Private Finance Initiative (PFI). The intention was to introduce the supposedly superior project management skills of the private sector into transport, defence, the NHS, education and other public services.

9.3 Privatization in Brazil

The Fernando Henrique Cardoso government copied, almost exactly, the British privatization program. For example, the National Telecommunications Agency (Anatel) was created in Brazil, as a copy of its British counterpart OFTEL, to regulate the competition among the privatized companies. The PPP has also been touted, since then, as a solution to privatization of sensitive sectors of the economy.

In spite of that, privatization was one of the reasons for the fixed and mobile telephony costs being among the highest in the world in Brazil, according to a report by the International Telecommunications Union (ITU) (Alencar, 2011b).

To get to the heart of the topic, it is worth remembering the work of Kurt Rudolf Mirow (1936–1992) and René Armand Dreifuss (1945–2003), who developed important theses for the understanding of the recent history of the country (Mirow, 1978) (Dreifuss, 1981).

Kurt Rudolf Mirow became famous when he was prosecuted according to the National Security Law by Minister Armando Falcão, when he published, in 1978, the book "The Dictatorship of Cartels", in which he denounced a plan of the multinationals to dominate the Brazilian market.

The book has an impressive testimony about how multinational companies operate. The author, an industrialist with no commitment to ideologies or political movements, was director of the company Codima S.A. and suffered the maneuvers and threats of an international cartel. Because of that, he began to investigate the problems that large corporations create by dominating the markets with harmful agreements for the countries where they operate.

The book was a huge success because it revealed, for the first time, in a a well documented publication, the obscure practices of large multinational companies, with regard to the conquest and maintenance of captive markets for its products, and the elimination of competitors. Mirow also published the sequels, entitled "Condemned to Underdevelopment" in 1978, and "Nuclear Madness" in 1979, a critique of the Brazilian nuclear power program (Mirow, 1978).

His thesis explains the cartelization of the Brazilian market in sectors that go from telecommunications to automobiles, in which companies combine prices and set them up to three times higher than those practiced in the United States, for example, discounted the taxes. In the area of telecommunications, prices can reach 200 times those charged in India, for example.

This formation of cartels became evident in the area of communications after the privatization. The actual prices of services, since then, have been strictly identical, even if disguised in promotions, bonuses, and other tricks used by the companies. With this combination of prices, companies have always accumulated stratospheric profits.

Because of Vivo's revenue, its holding company, Telefónica de Espanha, obtained, record profits in Brazil. This is an indicator of the immense shipment of profits made abroad, and this is just one of the companies that dominate the sector. It is interesting to note that, in Spain, Telefónica had a loss in the same period! At the headquarters of the multinationals, governments control, more or less, the formation of cartels (Alencar, 2011c).

René Armand Dreifuss published the book "1964: The conquest of the state. Action Politics, Power and Class Coup", in 1981, in which he analyzes the events of the 1964 *coup d'état*, as a carefully engineered project to seize power in Brazil. Dreifuss pointed out the participation of economically dominant sectors in the overthrow of president João Goulart (1919–1976) (Dreifuss, 1981).

He also showed the role played by civilian institutions in overthrowing the democratic government, such as the Women's Campaign for Democracy (CAMDE), the Brazilian Institute Democratic Action (IBAD) and the Institute for Research and Social Studies (IPES), in addition to the role of newspapers and magazines of great circulation which, at the time, supported the coup.

These periodicals, in addition to radio and television companies, received major sums, usually in the form of advertising funds, to create, in the middle class, the illusion that the country was about to be taken by the communists. And the effect had a devastating impact on this sector of the population, which joined the coup.

The founders of these bodies had different ideological profiles but shared a certain identity in their multinational and associated economic relations, in its anti-communist position and in their ambition to reform the State. Dreifuss added to the book a thorough documentary research to show the participation of the organized civil society, until then little perceived, in the action of the military that plotted the coup.

Companies were created, in Brazil and abroad, to collect money from entrepreneurs and invest it in projects such as the purchase of magazines and newspapers, investment in ads in all media, payment of talks by specialists and payment of tuition fees to train students. Politicians, who could not profit directly from deposits of those fake companies, received money to give lectures, for example.

Dreifuss' work illuminates the dark side of privatization, perpetrated in the Cardoso government, but that had begun at the time of José Sarney, and continued during the government of Fernando Collor de Mello.

For more than a decade, newspapers and magazines were recruited, with large sums of advertising money, to criticize the model of state-owned companies that was created by the military, which ruled Brazil, from 1964 to

1985. This state-owned model started with the creation of Embratel in 1965, but it was only fully implemented with the formation of Telecomunicações Brasileiras S.A. (Telebrás).

Since the Juscelino Kubitschek term in the Brazilian presidency (1902–1976), telecommunications was collapsing in the country. Private companies, which dominated the generation, transmission, and distribution of energy, transportation by trams, telecommunications, mining, among other areas, had long stopped investing in the Country Several companies operated in Brazil since the time of the Empire (Alencar, 2011d).

There was no adequate telephone service in Brazil, and whoever had the device did not hear a dial tone, because the exchanges were congested. The military, probably unwillingly, ended up taking a socializing measure and nationalized the companies. However, all the headquarters of the multinationals received large amounts of compensation money for the nationalization of their Brazilian subsidiaries.

Telebrás, the state-owned holding of the telecommunication companies, was only created in 1972, which shows that nationalization had an ephemeral life in Brazil. Because the state-owned companies have been privatized, in 1997, the period of state-owned companies lasted only 25, out of 521 years of Brazilian history. An illusion created by an official, and private, propaganda convinced the middle class that telecommunications had always been in the hands of the State.

It is interesting to note that, unlike almost all European countries, radio and television have always been private in Brazil. A model that president Getúlio Vargas (1882–1954) tried to change, with the nationalization of Rádio Nacional, created by Edgard Roquete-Pinto, in 1923, as Rádio Sociedade of Rio de Janeiro, but that did not succeed.

The paradox of all this is to note that nationalization, a typically socialist attitude, was done by the military, notoriously anti-communist. And privatization, an action generally attributed to the right-wing, was carried out by the socialist Fernando Henrique Cardoso, currently a famous speaker, who receives US$50,000 for each lecture given, and that, curiously, decided to defend the release of marijuana in Brazil.

In April 2004, less than two years after leaving the government, Cardoso had earned about US$600 thousand giving lectures in Brazil and abroad, and received more than US$3 million in donations from entrepreneurs for his institute.

And Brazil, which had its own patents and was self-sufficient in the production of almost all telecommunications equipment, from telephone exchanges to optical cables, became an importer of equipment. For instance, the trade deficit in the electronics sector reached US$5.46 billion, in 2010. That was US$1 billion more than the total amount collected by the Cardoso government with the sale of all telecommunications companies in the country!

In a kind of terrible nostalgia from the pre-nationalization era, today it is difficult to complete a call with a single attempt and keep the connection active throughout the conversation. Accesses are always busy, and connections are lost and redone, several times in just one conversation (Alencar, 2011e).

The lack of investment by private companies led the country to a state of near chaos, in which service fees for 4G are very expensive. On the other hand, the transmission rates, which should provide 100 Mbit/s, do not reach 2 Mbit/s in the outskirts of the main cities. That forced the federal government to revive Telebrás, to install a broadband network, with adequate transmission rates and costs.

On the other hand, blocking rates, which should not exceed 1%, for the fixed service, and 2% for the cellular mobile service, currently reach 10% and 20%, that is, one phone call in five is blocked, due to lack of links, during Busy Hour (BH).

Connection drops are frequent, mainly due to the indiscriminate use of voice over Internet protocol (VoIP) by companies, without concern with quality of service (QoS), which causes packet loss. Worst of all, the users are paying for the circuit type transmission service, theoretically more stable, and receiving the datagram service, or packet, which is more inefficient for voice traffic.

Former president Dilma Rousseff (1947-), who decided to privatize some state-owned companies, certainly had no nostalgia for the military era, nor for the pre-nationalization period, when most companies had an employee known as "boy at the telephone," who was responsible for the thankless task of taking the telephone off the hook, and waiting for the dial signal, when the boss needed to place a call. And this could take hours!

She was probably not old enough, at the time, to remember a pathology, classified as the telephone syndrome, reported in publications in the United States, which affected many Brazilian subscribers who were unable to place calls. That was the state of affairs at that time, before the creation of Embratel and Telebrás.

Certainly, she did not read the poet Carlos Drummond de Andrade's (1902–1987) chronicle, which tells the story of the unfortunate candidate to be a subscriber of the Brazilian Telephone Company (CTB), which despite the name was Canadian. The story goes explaining that he had received a promise to install a telephone line from his father, as an inheritance, but that he dies at the office of the CTB, before passing the contract to his child.

9.4 The Companies that Dominate the Telecommunications Sector in Brazil

Anatel evaluated the groups with Significant Market Power (SMP) in the country. Because of the control they have over these markets, these telecommunication operators must comply with specific regulatory measures, for instance, regarding the sharing of their networks, or how to maintain controlled reference prices (Alencar, 2011a).

In 2011, the infrastructure and broadband offer in Brazil counted approximately 1,200 municipalities, where there was no company with significant market power. Telemar had SMP in 3,763 municipalities, Telefônica controlled 591 cities, and Telmex 49. Telefônica and Telmex shared market power in 44 municipalities and Telmex and Telemar together controlled five cities.

Regarding fixed and mobile interconnection, the Telemar group had SMP in 4,919 municipalities and Telefônica in 645 municipalities. In local transport, Telemar maintained SMP in 3,934 cities, Telefônica in 606 municipalities and there were 1,025 municipalities without a single operator with significant market power.

In the long-distance communications market, Telemar had SMP in 3,934 counties. Telmex had SMP in 1093 municipalities and, in 687 of them, it shared the condition with Telemar. The Telefônica group controlled 606 municipalities, in 295 of which Telmex also exercised power. In 913 municipalities there was no operator with SMP.

Considering the infrastructure for accessing the mobile network, the telecommunications operator Oi had SMP in 25 registration areas, Vivo in 16, Claro in six, and TIM in one. Regarding the interconnection in mobile networks, Oi had SMP in the regions of its concession area and Vivo had SMP in São Paulo.

In markets where calls originate on the fixed telephone network, Oi and Telefônica were defined as operators with SMP in municipalities where they had more than 20% of the market. In the market for calls terminated on telephone networks, Telefônica and Telemar were considered to have SMP.

Finally, the telecommunication agency found that, in the pay TV market, only the Telmex group, which controls Net, had significant market power in the 94 municipalities in which it has a concession.

Brazil is a major exporter of commodities, mainly raw materials, such as iron and soybeans, whose prices have remained high due to the high consumption in China. In 2011, exports totaled US$235 billion, and imports, US$209 billion, with a positive balance in favor of Brazil, for the year, above US$28 billion (Alencar, 2012a).

However, the largest item on the Brazilian export basket is a huge loss to the country. Brazil continues to be a major exporter of wealth. In particular, companies in the telecommunications sector privatized under the Fernando Cardoso government, export a huge amount of money to their headquarters abroad.

The remittance of funds abroad follows the usual profile, which involves overpricing of purchases, indebtedness of the company to the parent company, or sending cash in suitcases, with each carrier taking, on average, US$500 thousand per trip, on the pretext of participating in professional events, promoted by the companies. The priests are not the only ones who travel with money in their suitcases, underwear and bibles!

9.5 Lack of Investment Paralyzed Brazil

Probably the most familiar diagram in the telecommunications economy is the one that shows the relationship between phone density and per capita income of a country. In a study published in 1963, engineer and professor A. G. W. Jipp demonstrated a positive correlation between a nation's income and telephone density (Alencar, 2013c).

This relationship became known as Jipp's Law and indicates that an increase of a thousand dollars in the GDP per inhabitant is associated with a 2.4-fold increase in tele-density. Generally, the relationship gets stronger as wealth increases. This is often interpreted as having a causal relationship that works in two directions: better communication generates economic growth, which demands a better communications infrastructure. (Alencar, 2018d).

The researcher Márcio Wohlers also observed in his thesis, in 1994, that there was a very strong correlation between these two basic indicators, used in the econometric study made by Jipp, which was endorsed by later studies. The correlation found between those indicators is significant, either in relation to the temporal evolution, in the internal affairs of a country, either by means of a comparison between a group of countries. He also observed a logarithmic relationship between the telephones *per capita* and GDP per capita variables.

The task to identify which is the dependent variable and which is the independent one, in this correlation, is an essential issue and has intrigued researchers for five decades. Wohlers concluded that econometric analysis did not allow an adequate response, and, therefore, it was not possible to establish causal relationships between the telecommunications density and the economic development, as measured by *per capita* income.

However, the size of infrastructure networks, in general, and not only telecommunications, reflects the level of development of a country. The network size is also one of the indicators of the country's level of development. He pondered that, if it were possible to prove that the expansion of the telecommunications network has a more important relative effect than that of the other infrastructures, to leverage the economic development, there would be greater technical and theoretical consistency in proposals to increase the resources to be allocated to this sector.

The impact of a telecommunications system on society stems from its ability to shorten distances, decreasing cost, and reducing the necessary time for the transmission of information. Regarding personal motivation, telecommunications provide a means of interactive and immediate communication. The use of telecommunications for personal purposes has low intensity, and there is a considerable idle capacity of the equipment that is exclusively allocated for this type of use.

The commercial use, to do business, on the other hand, rely on the same characteristics of instantaneity and interactivity provided by telecommunications, but has a strong impact on the internal operation of firms and companies, which is an important factor for the organization of markets and also for the economic development.

In addition to supporting transactions, which allows for an improvement in microeconomics efficiency and the organization of markets, telecommunications still presents a third impact, related to the spatial or regional dimension of economic development, because the telecommunications infrastructure acts in a complementary way, or possibly substitutive, in relation to other physical transport infrastructures.

This is an important point, because the complementary, or even substitutive, action of communications in relation to the transport infrastructure, for example, indicates that there may be, after all, a sense of cause and effect in Jipp's Law.

As is well known, the construction of highways, bridges and other transport and logistics structures has the potential to induce the development of the region where the investment is made. There is a notorious pressure from governments to obtain resources for road construction, due to this inducing effect economic growth.

Because the telecommunications infrastructure is the information highway, related to the traffic of videos, messages, images, data, packages, transactions among others, and replaces or complements other infrastructures, it is natural to imagine that it is also an inducer of economic growth.

In this way, it is easier to understand why Brazil, for example, had its growth stalled in recent years: the lack of investment on the part of the communications operating companies is certainly the main reason. Operators are more concerned with repatriating, or expatriating, the profits obtained from the high-level prices that they charge in the country. They did not make the necessary investments to improve the telecommunications services. The voice and data services are poorly provided.

To complete a phone call it is necessary to try several times, but a call drop in the middle of the conversation is common, and the blocking probability is ten to twenty times that specified by Anatel. On the other hand, the effective transmission rate never reaches 20% of that contracted, configuring a misappropriation of subscribers' resources, and obliging the federal government to resurrect Telebrás to provide the well-known broadband services, because the privatized companies did not make the necessary investments.

Jipp's Law also indicates that, for Brazil to reach the level of a developed country, it is necessary to make huge investments in telecommunications, mainly in the communications network, in the telephone exchanges, in the switching and control centers, in base stations, in broadcasting, in computer networks, and not only sell cellphones.

9.6 About Privatizations, Service Charges and Taxes

The privatization of state-owned companies, in addition to the evident transfer of public resources from the State to private companies, the poor quality of services provided by the companies that purchased them, and the high cost of these services for the users also caused a serious problem for taxpayers. A little macroeconomic theory helps to understand the problem (Alencar, 2013e).

Governments have three main ways of financing the development of their countries, to carry out public works and apply health and education resources, among other priorities. The first and most common is indebtedness, which can be done with banks national or international, and also by issuing papers, like treasure bills.

The second way of obtaining revenue is by collecting taxes, an inefficient process that tends to burden small traders and low- and middle-income taxpayers, but mainly civil servants and employees of large companies, which are the most targeted by the tax authorities. Furthermore, it is subject to evasion and criticism from all sectors, because it is a sort of billing that, usually, has no defined application.

The third way of financing is by billing, made by state-owned companies. This is the best, fairer, and more efficient way to earn revenue because it works as a tax with a specific destination, paid by the person who actually uses the service. The population does not complain about the payment of bills or evade it when one realizes that it is fair in relation to the service provided.

With the sale of state-owned companies, the government gave up the charging bills in favor of foreign entrepreneurs and capitalists, or not so much.

Because the government of Fernando Henrique Cardoso, also known by the nickname FHC, had no credibility to launch additional papers or take new loans, it started to charge more taxes.

The public debt, at the beginning of the FHC government, was 30% of the GDP, the sum of the whole country wealth. The tax burden was 27% of the GDP. At the end of the FHC government in 2002, after the privatization of several state-owned companies, according to the Central Bank (BC), the gross debt of the federal government amounted to 62.7% of the GDP, the net debt reached 60.4%, and the taxes on the population rose to 36% of the GDP.

In 1997, the government raised US$21 billion in taxes. At the end of the FHC administration, in 2002, the nominal value had passed US$122 billion, that is, almost 100% increase in taxes levied, compared to the year the sale of state-owned companies was approved. It is worth remembering that one of the objectives alleged for privatization was exactly to reduce the burden tax levied on the taxpayer.

Interestingly, the biggest increase in tax collection in the recent history of Brazil occurred exactly between 1997, the year the Constitution was changed to allow the breaking of the monopoly in various services, and 1998, the year the last state-owned companies, in the telecommunication sector, were sold.

Taxes increased by no less than 13.85%, at prices of December 1998. What resulted from the exchange made by the FHC government, failing to collect the bills by the state-owned companies, and starting to collect more taxes, due to the sale of these companies. But, the population just realized that problem a few years after the privatization took place.

The problems associated with privatization do not stop at the numbers cited in the preceding. The Brazilian international reserves fell, at the end of the Fernando Cardoso government, to the lowest level in decades, only US$38 billion. Even after the government had privatized all telecommunication companies (Alencar, 2013d) (Dória, 2011).

To provide a comparison, the international currency reserves, in 2012, during the Dilma Rousseff government, were US$378.9 billion, that is, 900% higher than those of the time the privatizations took place.

The external debt, which should have been reduced with the privatizations, as promised by José Serra, ex-finance minister who planned and coordinated the process, reached, in 2002, the highest value in the history of the country, nothing less than US$211.7 billion, according to data published by the Brazilian Central Bank.

Furthermore, the unemployment rate, fueled by mass layoffs made by the consortia that acquired the state-owned companies, rose to 12.5%, one of the largest in the history of Brazil. Inflation, the only indicator that the FHC government claimed to have improved, reached, in 2002, an impressive 12.53%, the highest value in the last two decades (Jr., 2011).

Finally, the GDP growth, which is a good indicator of success or failure of the privatization process, was only 0.3%, in 1999, the year following the privatization. In order to provide a comparison, the average GDP growth in the decade that followed the FHC government, including the year 2013, was 3.9%, more than thirteen times the average value in the FHC era.

Brazil ended FHC's term, which fulfilled all the premises of world neoliberalism, including the sale of the State assets for insignificant values, even inferior to the own annual turnover of the same companies, classified as speculative grade by all risk assessment agencies worldwide.

During the governments of Luiz Inácio Lula da Silva (1945–) and Dilma Vana Rousseff (1947–), in comparative terms, the country did not sell essential public companies, and was classified as investment grade by all agencies. The long-term local currency debt, which represents more than 95% of the federal government debt, was classified as A- by Standard & Poors (S&P), with stable perspective.

Figure 9.1 shows a portrait of President Lula, who ruled Brazil for two terms.

The service provided by the telecommunication operators and by electric power facilities, in Brazil, is of very poor quality, and the prices paid by

Figure 9.1 Luiz Inácio Lula da Silva ruled Brazil for two terms. Image adapted from Wikimedia Commons, a collection of free content from the Wikimedia Foundation.

users are one of the highest in the world, according, for example, to the International Telecommunication Union (ITU).

The Vale do Rio Doce company, for instance, the largest diversified mining company in the Americas and the second largest in the world, also privatized during the FHC government, for a price well below its market value, did not meet the goal of building a steel plant in the Country and continues to pay its shareholders mainly with the export of iron ore.

Interestingly, the Vale do Rio Doce company, whose market value, in 2011, was US$190 billion, with a turnover of US$100 billion and a profit of US$41.4 billion per year, had its share control sold for only US$1.65 billion in the FHC government.

And Brazil, 521 years after the discovery, continues to export ores, at the price of brazilwood, and buy manufactured goods at absurd prices established by the current colonizers. In addition, the Country continues to export capital through privatized companies, via overpricing of purchases abroad, sending money in suitcases, alienation of assets, among other ways to circumvent the tax authorities.

9.7 Whom Privatizations Served?

The General Telecommunications Law was enacted by then President Fernando Henrique Cardoso, in 1997. This Law altered the state monopoly regime in the sector, to allow the privatization of public companies and the sale of new licenses for the provision of services (Alencar, 2007).

The process, much criticized until today, passed the control of almost all companies that belonged to the Telebrás holding to foreigners investors at very low prices. The investment made by the government in state enterprises between 1993 and 1997 was higher than the total collected from the sale of assets and licenses by Anatel.

Few companies were left in the hands of Brazilians, such as Telemar who stayed with the family of Tarso Jereissati from Ceará State. The bank Opportunity, owned by Daniel Dantas, linked to the deceased governor Antônio Carlos Magalhães, controlled several companies, despite the lack of experience in the area. Later on, Daniel Dantas was involved in a fraud investigation by the Congress (Vaz, 2012).

Embratel, which had been sold, in 1998, for a measly US$650 million, had a turnover of almost US$5 billion, in 2007, and contributed to positioning the Mexican Carlos Slim as the richest man on Earth. At the time, the company profited US$1 billion, and was one of the most solid companies in the area (Alencar, 2012c).

All over the world, and especially in Brazil, privatizations of State-owned companies have been associated with scams, schemes for illicit enrichment, favoring friends and supporters, or support for political campaigns (Alencar, 2012e).

Apparently, the investors are never interested in putting their own resources in new ventures to compete with state-owned companies, but do not shy away from going to government auctions to acquire companies, as long as prices are low.

Public-private partnerships are also examples of businesses in which, generally, the government provides the resources, the area, the improvements, the tax exemptions, and entrepreneurs just keep the profits, typically in the form of tolls, boarding fees, service charges, or tariffs.

Evidently, oligopolies are formed that expropriate the population without providing adequate service, as occurs today, for example, in the areas of telecommunications and electricity, until a future government revokes the concessions. And the cycle repeated itself, for centuries.

9.8 Anatel Played the Operator's Game

One of the problems associated with the selling strategic sectors of a country to private companies is the sharp loss of revenue. First, the government loses the revenue that comes from service charging, when it sells the state-owned companies (Alencar, 2014a).

This revenue is never offset by tax collection, as the neo-liberals claimed when they articulated the auctions of privatization because companies prefer to pay lawyers and accountants, to circumvent the payment of most taxes, and fool the tax authorities.

To this end, entrepreneurs change the shareholding composition of companies, change their corporate names, exchange assets between them, sell assets, which were never mentioned at the time of the auctions, rent buildings to simulate increased costs and lower revenues, a multitude of accounting and legal tricks to deceive the fiscal authorities.

The country, of course, never makes any money with its own privatization auctions, because the companies are usually sold for very low prices, usually only a small fraction of their market values. This is, certainly, agreed upon between the parties beforehand. Of course, everything is financed by official banks.

But, the bleeding did not stop in these lurid operations, in the case of telecommunication companies, there are also auctions of operating frequencies. These auctions yield a lot of money to governments abroad, tens to hundreds of billions of dollars. In some countries, for instance, it is never the case, because the frequency lots are just given away by the governments.

10

Mathematical Economy

"If a free society cannot help the many who are poor, it cannot save the few who are rich."
John F. Kennedy

10.1 Introduction

This chapter presents the main economic definitions and the mathematical tools that are useful to understand the subject of Mathematical Economy, from a stochastic process point of view. The Glossary on Economy, available as an appendix, contains the terminology associated with the area that could be helpful to the reader.

10.2 Econometrics

Econometrics is a branch of Economy that aims at measuring and estimating the relation between two or more economic variable. It combines economic theory, mathematical economy, computing science, statistics, and stochastic processes, with the purpose of providing numerical values to economic relations, such as marginal values, means, and trends, to verify economic theories (Koutsoyannis, 1977).

One of the most important features of economic relations is the randomness of some elements. and Econometrics developed, or adapted, methods to deal with such random components. The researchers in the area search for estimators that have desired statistical properties, such as, efficiency, consistency, and unbiasedness.

Jan Tinbergen (1903–1994) an important Dutch economist, who received the Nobel Prize in Economy, is considered one of the founders of Econometrics, and has developed the first macroeconomic models. He also understood the dynamic models and obtained a solution to the identification problem.

The economic data in the econometric empirical analysis are obtained from the observation of economic and social phenomena, and can be classified as:

- Quantitative data, when they assume numerical values, for example, monetary values, volume measurements, extension, or proportion.
- Qualitative data, when they assume values that are expressed as attributes, such as, sex, residence, or formation.

Regarding the format that they take, the data used in estimating a certain model can correspond to the following characteristics: (Koutsoyannis, 1977):

- Data from time series are observations about one or several variables along time. They usually refer to annual intervals, trimesters, months, and days. A few examples of time series are: gross national product, consumer price index, and automobile sales.
- Transversal cohort data (transversal sections data) consist of a sample of individual, consumers, companies, cities, states, countries, or a variety of other units, taken at a certain point in time.
- Grouped transversal cohort form a combination of data of grouped cross sections and time series. For instance, when two samples of data from an economic research are used, in different years, possibly with distinct subjects.
- Panel data, or longitudinal, consist of a time series for each member of the transversal section of the set of data. For example, the financial and investment collection of data from a set of companies over a certain period of years. In this case, the same companies are accompanied for a determined time.
- Engineering data, that provide information on the technical requirements of the production method employed to deliver a certain good. The data are obtained from the producers and are used in studies about the production.
- Legislation and regulations, that can provide information on the nature of the involved relations, such as, in the case of taxation.

Econometrics turned into an important tool to improve most economic models, and to the development of public policies. Its main objective is hypothesis testing and simulation of cases, in order to find the optimal solution to problems faced by society. To attain such goals, the following scheme is used (Koutsoyannis, 1977):

1. Data research, collection, and measurement, to obtain the model variables;

2. Parameter definition, that simplifies the economical reality;
3. Hypothesis proposition and test;
4. Obtaining estimates for expected results.

The analysis and quantification of the studied models allows attaining the following objectives:

- Verify the validity of economic theories, by means of statistical tests, that are necessary to prove, in a scientific manner, the presented hypotheses;
- Use the studied methods, such as the linear regression, to forecast future economical trends.

Econometrics makes it possible to estimate economic relations, in macroeconomic, as well as, in macroeconomic; test economic theories; evaluate the implementation of government policies, using, for instance, macroeconomic variables, such as interest rate and inflation rate; implement managerial and business policies; in addition to assisting decision-making business or public policy.

10.3 Stochastic Modelling of the Productive Process

Usually, Statistical Process Control (SPC) is the preferred tool to characterize common and specific events that disturb the production, or the general economy. But, because the analysis is based on Statistics, that is, it does not involve time, it does not allow inference or trend forecasting, as is the case in most economical problems, for instance, in the prevision of reject from an industry.

The Theory of Stochastic Processes (TSP) is a powerful tool to study random time series, and permits to verify the correlation between production factors, as well as, the forecasting of future events, or their modeling.

10.4 Statistical Correlation Analysis

The usual solution to problems in the economy uses statistical correlation, a tool that derives from the study of stochastic processes (Alencar, 2014h).

A random process, or a stochastic process, is an extension of the concept of a random variable, involving a sample space, a set of signals, and the associated probability density functions. Figure 10.1 illustrates a random signal and its associated probability density function.

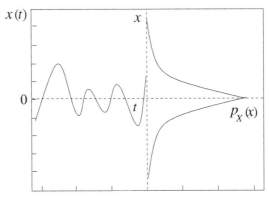

Figure 10.1   Example of a random process and its corresponding probability density function.

A random process (or stochastic process) $X(t)$ defines a random variable for each point on the time axis. A stochastic process is said to be stationary if the probability densities associated with the process are time independent.

10.4.1 The Autocorrelation Function

An important joint moment of the random process $X(t)$ is the autocorrelation function

$$R_X(\xi, \eta) = E[X(\xi)X(\eta)], \tag{10.1}$$

in which

$$E[X(\xi)X(\eta)] = \int_{-\infty}^{+\infty} \int_{-\infty}^{+\infty} x(\xi)x(\eta)p_{X(\xi)X(\eta)}(x(\xi)x(\eta))dx(\xi)dy(\eta) \tag{10.2}$$

denotes the joint moment of the r.v. $X(t)$ at $t = \xi$ and at $t = \eta$.

The random process is called wide sense stationary if its autocorrelation depends only on the interval of time separating $X(\xi)$ and $X(\eta)$, i.e., depends only on $\tau = \xi - \eta$. Equation 10.1 in this case can be written as

$$R_X(\tau) = E[X(t)X(t + \tau)]. \tag{10.3}$$

10.4.2 Stationarity

In general, the statistical mean of a time signal is a function of time. Thus, the mean value

$$E[X(t)] = m_X(t),$$

the power

$$E[X^2(t)] = P_X(t)$$

and the autocorrelation

$$R_X(\tau, t) = E[X(t)X(t + \tau)],$$

are, in general, time dependent. However, there exists a set of time signals the mean value of which are time independent. These signals are called stationary signals, as illustrated in the following example.

Example: Calculate the power (RMS square value) of the random signal $X(t) = V\cos(\omega t + \phi)$, in which V is a constant and in which the phase ϕ is a r.v. with a uniform probability distribution over the interval $[0, 2\pi]$. Applying the definition of power it follows that

$$E[X^2(t)] = \int_{-\infty}^{\infty} X^2(t)p_X(x)\,dx = \frac{1}{2\pi}\int_0^{2\pi} V^2\cos^2(\omega t + \phi)\,d\phi.$$

Recalling that $\cos^2\theta = \frac{1}{2} + \frac{1}{2}\cos 2\theta$, it follows that

$$E[X^2(t)] = \frac{V^2}{4\pi}\int_0^{2\pi}(1 + \cos(2\omega t + 2\phi))\,d\phi = \frac{V^2}{4\pi}\phi\Big|_0^{2\pi} = \frac{V^2}{2}.$$

Since the mean value m_X of $X(t)$ is zero, i.e.,

$$m_X = E[X(t)] = E[V\cos(\omega t + \phi)] = 0,$$

the variance, or AC power, becomes

$$V[X] = E[(X - m_X)^2] = E[X^2 - 2Xm_X + m_X^2] = E[X^2] = \frac{V^2}{2}.$$

Therefore,

$$\sigma_X = \frac{V}{\sqrt{2}} = \frac{V\sqrt{2}}{2}.$$

Application: The dynamic range of a stochastic process, from a probabilistic point of view, is illustrated in Figure 10.2. As can be seen, the dynamic

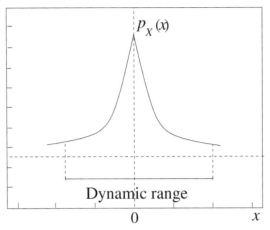

Figure 10.2   Dynamic range of a stochastic process.

range depends on the standard deviation, that is usually specified for $2\sigma_X$ or $4\sigma_X$. For a process with a Gaussian probability distribution of amplitudes, this corresponds to a range encompassing, respectively, 97% and $99,7\%$ of all signal amplitudes.

However, since the process is time varying, its statistical mean can also change with time, as illustrated in Figure 10.3. In the example considered, the variance is initially diminishing with time and later it is growing with time. In this case, an adjustment in the signal variance, by means of an automatic gain control mechanism, can remove the pdf dependency on time.

A stochastic process is stationary whenever its pdf is time independent, i.e., whenever $p_X(x,t) = p_X(x)$, as illustrated in Figure 10.4.

Stationarity may occur in various instances:

1. Stationary mean $\Rightarrow$ $m_X(t) = m_X$;
2. Stationary power $\Rightarrow$ $P_X(t) = P_X$;
3. First-order stationarity implies that the first-order moment is also time independent;
4. Second-order stationarity implies that the second-order moments are also time independent;
5. Narrow sense stationarity implies that the signal is stationary for all orders, i.e., $p_{X_1\cdots X_M}(x_1,\cdots,x_M;t) = p_{X_1\cdots X_M}(x_1,\cdots,x_M)$

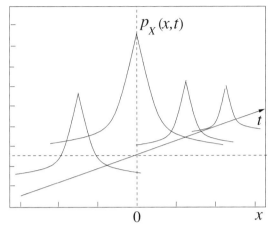

Figure 10.3 Time varying probability density function.

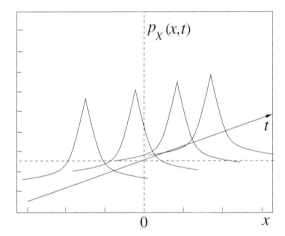

Figure 10.4 Time invariant probability density function.

10.4.2.1 Wide Sense Stationarity

The following conditions are necessary to guarantee that a stochastic process is wide sense stationary.

1. The autocorrelation is time independent;
2. The mean and the power are constant;
3. $R_X(t_1, t_2) = R_X(t_2 - t_1) = R_X(\tau)$. The autocorrelation depends on the time interval and not on the origin of the time interval.

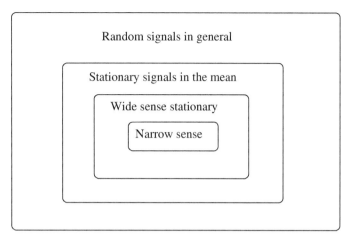

Figure 10.5   Stationarity degrees.

10.4.2.2 Stationarity Degrees

A pictorial classification of degrees of stationarity is illustrated in Figure 10.5. The universal set includes all signals in general. As subsets of the universal set there are the signals with a stationary mean, wide sense stationary signals, and narrow sense stationary signals.

10.4.2.3 Ergodic Processes

Ergodicity is another characteristic of random processes. A given expected value of a function of the pdf is ergodic if the time expected value coincides with the statistical expected value. Thus ergodicity can occur on the mean, on the power, on the autocorrelation, or with respect to other quantities. Ergodicity of the mean implies that the time average is equivalent to the statistical process average. Therefore,

- Ergodicity of the mean: $\overline{X(t)} \sim E[X(t)]$;
- Ergodicity of the second moment: $\overline{X^2(t)} \sim \overline{R_X}(\tau) \sim R_X(\tau)$;
- Ergodicity of the autocorrelation: $\overline{R_X}(\tau) \sim R_X(\tau)$.

A strictly stationary stochastic process has time independent joint pdf's of all orders. A stationary process of second-order is that process for which all means are constant and the autocorrelation depends only on the measurement time interval.

Summarizing, a stochastic process is ergodic whenever its statistical means, which are functions of time, can be approximated by their corresponding time averages, which are random processes, with a standard deviation that is close to zero. The ergodicity may appear only on the mean value of the process, in which case the process is said to be ergodic on the mean.

10.4.3 Properties of the Autocorrelation

The autocorrelation function has some important properties as follows.

1. $R_X(0) = E[X^2(t)] = P_X$, (Second moment, or total power);
2. $R_X(\infty) = \lim_{\tau \to \infty} R_X(\tau) = \lim_{\tau \to \infty} E[X(t + \tau)X(t)] = E^2[X(t)]$, (Average power);
3. Autocovariance: $C_X(\tau) = R_X(\tau) - E^2[X(t)]$;
4. Variance: $V[X(t)] = E[(X(t) - E[X(t)])^2] = E[X^2(t)] - E^2[X(t)]$ or $P_{AC}(0) = R_X(0) - R_X(\infty)$;
5. $R_X(0) \geq |R_X(\tau)|$, (Maximum at the origin);
   This property is demonstrated by considering the following tautology

$$E[(X(t) - X(t + \tau))^2] \geq 0.$$

Thus,

$$E[X^2(t) - 2X(t)X(t + \tau)] + E[X^2(t + \tau)] \geq 0,$$

i.e.,

$$2R_X(0) - 2R_X(\tau) \geq 0 \implies R_X(0) \geq R_X(\tau).$$

6. Symmetry: $R_X(\tau) = R_X(-\tau)$;
   In order to prove this property it is sufficient to use the definition $R_X(-\tau) = E[X(t)X(t - \tau)]$
   Letting $t - \tau = \sigma \implies t = \sigma + \tau$

$$R_X(-\tau) = E[X(\sigma + \tau) \cdot X(\sigma)] = R_X(\tau).$$

7. $E[X(t)] = \sqrt{R_X(\infty)}$, (Process mean value).

Application: The relationship between the autocorrelation and various measures is illustrated in Figure 10.6.

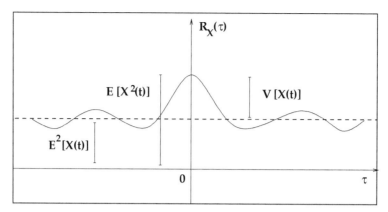

Figure 10.6   Relationship between the autocorrelation and various measures.

10.4.4 Model for the Prediction of Random Series

Given a stationary random process $X(t)$, it is possible to estimate its future value $X(t+\tau)$, that is, τ units ahead in time, if one can compute the processo autocorrelation $R_X(\tau)$ and its derivative.

Suppose a linear prediction, of the proportional and derivative type,

$$X(t+\tau) \approx \alpha X(t) + \beta X'(t), \tag{10.4}$$

in which α and β are parameters to be determined, that depend on the original process, and $X'(t)$ represent the derivative of process $X(t)$.

It is possible, using the minimization of the mean square error, to determine the parameters as a function of the process $X(t)$ autocorrelation. The mean square error between the process and its future value is given by

$$\epsilon = \mathrm{E}[(X(t+\tau) - \alpha X(t) - \beta X'(t))^2],$$

The error minimization procedure requires the partial derivatives of the error in relation to the adjusted parameters α and β. Those derivatives are set to zero, that is

$$\frac{\partial \epsilon}{\partial \alpha} = \frac{\partial}{\partial \alpha} \mathrm{E}[(X(t+\tau) - \alpha X(t) - \beta X'(t))^2] = 0$$

and

$$\frac{\partial \epsilon}{\partial \beta} = \frac{\partial}{\partial \beta} \mathrm{E}[(X(t+\tau) - \alpha X(t) - \beta X'(t))^2] = 0.$$

Taking the partial derivatives, considering that the expectancy operator is linear, one can obtain the following system of linear equations:

$$E[-2(X(t+\tau) - \alpha X(t) - \beta X'(t)) \cdot X(t)] = 0$$

$$E[-2(X(t+\tau) - \alpha X(t) - \beta X'(t)) \cdot X'(t)] = 0,$$

that can be written, using the definitions of correlation and autocorrelation, as

$$R_{XX}(\tau) - \alpha R_{XX}(0) - \beta R_{XX'}(0) = 0$$

and

$$R_{X'X}(\tau) - \alpha R_{XX'}(0) - \beta R_{X'X'}(0) = 0.$$

Using the properties of the correlation and autocorrelation functions, and solving the system of equations, it is possible to obtain the optimum values for the parameters of the formula:

$$\alpha = \frac{R_X(\tau)}{R_X(0)}$$

and

$$\beta = \frac{R'_X(\tau)}{R''_X(0)}.$$

Therefore, the final formula for the prediction of the stochastic process is given by

$$X(t+\tau) \approx \frac{R_X(\tau)}{R_X(0)} X(t) + \frac{R'_X(\tau)}{R''_X(0)} X'(t). \tag{10.5}$$

Thus, for a short Tina interval, it is possible to forecast any economic series, based on its current value $X(t)$ and on its first derivative $X'(t)$, using the process autocorrelation $R_X(\tau)$ and its first and second derivatives.

The autocorrelation is computed from the time series provided by the company. Figure 10.7 shows a schematic diagram of a linear predictor, based on the mean square error minimization.

10.5 Study of Losses in the Production Line of a Large Footwear Company

This section analyzes the production loss, also referred to as reject, in a large company. This section presents a discussion of the probabilistic modeling, the development of a stochastic model, that uses Itô stochastic integration, based on temporal series provided by the company, and the identification of eventual solutions to the loss of production.

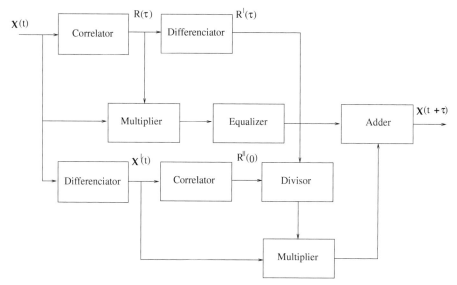

Figure 10.7   Schematic diagram of a linear predictor.

10.5.1 Introduction

This section discusses the problem of loss of production, also known as reject. The following sections present a discussion of the probabilistic modeling, and the development of a stochastic model, that uses Itô stochastic integration. The obtained results are based on the temporal series provided by the company. The section identifies eventual solutions to the large reject generated by the company and proposes a diagnostic to the production problem.

A discussion is presented that relates the proposed method to the classical Statistical Process Control (SPC) of production.

10.5.2 Production System

The study case refers to the Alpargatas Company, which started production in 1907. Its business has risen due to the growth of sales, and consequent increase in production and quarters, and the company is now the largest producer of footwear in the World. The Havaianas sandals, a World success, were created in 1962, with one model and a mix of four colors. Today there are 68 models for the national market, and 18 exclusive models for export with a large pallet of colors.

Havaianas are 80% of the rubber sandals sold in the country. Alpargatas is the leading company in the market in the footwear sector. The Campina Grande unit, different from the other companies of the group, does not have satellite factories. This industrial complex, formed by four factories and two Distribution Centers (DC), works in three shifts, from Monday to Saturday (Utsch and Letiere, 2014).

The Production Planning and Control (PPC) has been adopted to respond to market changes, and allows a continuous flow of materials, people, and information, able to feed this system, in order that the productive system meets the organization's strategic goals (Jaqueline Guimarães Santos, 2013).

The production averages 90% of its total capacity, which is 700.000 pairs per day. The creation of products is done in the São Paulo headquarters, and the development in the laboratories of the Campina Grande plant. The sandals are produced through an industrialization process, in which, 40% are residuals of the sandals, and 60% chemical products and input material. The productive flux has several steps in which the raw materials are transformed into compounds, all of these processes involve human interference, to put and remove loads, and make adjustments in the instrumentation of the control equipment (Utsch, 2014).

The formulation absorbs part of the reject and scraps of previous processes (40%), with errors (raw materials, manpower, instrumentation) from those batches that are now present in the formulation, in which new virgin (prime) raw materials, that were not used, are going to be mixed into.

When the compound starts a new cycle, it already has a certain amount of reject incorporated into it, from the previous process. After several generations of production, the reject takes on an average value, as in its formation it is composed of residuals from different products with separate reject (Soares et al., 2011).

This means that there is an initial value of the loss at each start of the process, and at the end, the other residuals are added to it, which can reach up to 20%, depending on the degree of difficulty in producing the sandal, and the amount of time to set the process parameters, as in the case of products from a new collection. To define the scope of the problem, Alpargatas has provided some files with information about the company's waste.

10.5.3 Performance Comparison

In the research, a study of the literature and the existing procedures was performed, to serve as a basis for the modeling of the random temporal series which characterizes the reject of the company. Generally speaking, the SPC is used to characterize common events, and specific events which disturb the production (Jaqueline Guimarães Santos, 2013) (Soares et al., 2011).

However, as the procedure is based on statistics, it does not involve the time variable, and does not permit an inference or prediction of tendencies for the reject. The Theory of Stochastic Processes (TSP) is a powerful tool for the study of random temporal series, and permits the verification of the correlation between production factors, as well, as the prediction of future events, or their modeling.

10.5.4 Study of Statistical Correlation

The aim of the study was to find a solution for the reject problem using statistical correlation. This tool results from the study of stochastic processes.

A stochastic or random process is an extension of the random variable concept, which comprises the sample space, the set of processes, and the associated probability density functions. In other words, a stochastic process is a family of random variables $\{X(t), t \in T\}$ defined on a probability space, and indexed by a time parameter t, which varies in a set T (Alencar and Alencar, 2016) (Alencar and da Rocha Jr., 2005).

10.5.5 Analysis Using Stochastic Integration

The stochastic processes are usually non-stationary, which means that their statistical averages, or moments, also vary with time. This makes it difficult to use correlation to treat them. In this case, stochastic integration is a useful tool to attack the problem, to obtain a problem formulation, from the modeling of the reject based on a stochastic differential equation.

The stochastic calculus began with the study and modeling of market prices, this is, the fluctuation of the stock value as a function of time. In this case, the investors work based on the variation of the potential gain or loss, $dX(t)$, as a proportion of the invested sum $X(t)$.

Therefore, in fact, what matters is the relative price, $dX(t)/X(t)$, of a certain asset, as it reacts to the market fluctuations, this is, it should be proportional to a Wiener process $W(t)$, (Brzezniak and Zastawniak, 2006).

$$dX(t) = \alpha X(t) dW(t),$$

which is an informal manner to express the corresponding integral equation,

$$X(t + \tau) - X(t) = \alpha \int_t^{t+\tau} X(u)dW(u). \tag{10.6}$$

An immediate question associated with the equation solution is related to the non-differentiability of a Wiener process $W(t)$, at any point in time. A way to circle the problem has been found and is known as the theory of stochastic integrals, or the study of stochastic differential equations (Itô, 1944).

The Itô general stochastic equation is given by

$$dX(t) = a(X(t), t))dt + b(X(t), t))dW(t), \tag{10.7}$$

in which $a(X(t), t))$ the drift function, or model trend, and $b(X(t, t))$ is the dispersion function, volatility, of the stochastic process.

10.5.6 Stochastic Model for the Reject Problem

When formulating a stochastic model to represent the reject variation along with time, it is important to consider that the loss rate is proportional to the existing quantity. This can be expressed as (Movellan, 2011).

$$\frac{dX(t)}{dt} = \alpha X(t). \tag{10.8}$$

The incremental variation of the reject is proportional to the differential variation $dW(t)$ of a stochastic process, $W(t)$, which usually has a Gaussian distribution, multiplied by the existing quantity of material, $X(t)$, and adjusted by the parameter β, which remains to be found, based on the production specifications

$$dX(t) = \beta X(t)dW(t). \tag{10.9}$$

The process $W(t)$ is the result of a combination of all production errors, $W_i(t)$, which can be found in the industrial production line

$$W(t) = \sum_{i=1}^{N} W_i(t), \tag{10.10}$$

thus, by the Central Limit Theorem, $W(t)$ has a Gaussian probability distribution, with mean and variance given by:

$$m_W(t) = E[W(t)] = \sum_{i=1}^{N} E[W_i(t)], \tag{10.11}$$

$$\sigma_W^2(t) = V[W(t)] = \sum_{i=1}^{N} V[W_i(t)]. \tag{10.12}$$

For stationary stochastic processes, the moments are independent of time, this is, $m_W(t) = m_W$ e $\sigma_W^2(t) = \sigma_W^2$.

Combining both assumptions, results in the following stochastic differential equation (Reiβ, 2007),

$$dX(t) = \alpha X(t)dt + \beta X(t)dW(t), \tag{10.13}$$

which is the informal manner to write

$$X(t+\tau) - X(t) = \alpha \int_t^{t+\tau} X(u)du + \beta \int_t^{t+\tau} X(u)dW(u), \tag{10.14}$$

in which $X(T+\tau)$ represents the future value of the process $X(t)$, that is τ time units in advance.

In order to solve the stochastic differential equation obtained with the reject model, using the generic formulation for stochastic differential equations, one must consider that the drift function is proportional to the stochastic process, this is, $a(X(t), t) = \alpha X(t)$, that the dispersion function is modeled as $b(X(t), t) = \beta X(t)$, and that one can use the Itô formula for the logarithm function $f(x, t) = \log(x)$ (Itô, 1946).

Following the procedure established by Itô, for the solution of the equation, one obtains,

$$\frac{\partial f(x, t)}{\partial t} = 0, \tag{10.15}$$

$$\frac{\partial f(x, t)}{\partial x} = \frac{1}{x}, \tag{10.16}$$

$$\frac{\partial^2 f(x, t)}{\partial^2 x} = -\frac{1}{x^2}. \tag{10.17}$$

Thus,

$$d\log(X(t)) = \frac{\alpha X(t)dt}{X(t)} + \frac{\beta X(t)dW(t)}{X(t)} - \frac{\beta^2 X^2(t)dt}{2X^2(t)}, \tag{10.18}$$

this is,

$$d\log(X(t)) = \left(\alpha - \frac{\beta^2}{2}\right)dt + \beta dW(t). \tag{10.19}$$

Integrating in time, one obtains

$$\log(X(t)) = \log(X(0)) + \left(\alpha - \frac{\beta^2}{2}\right)t + \beta W(t), \tag{10.20}$$

taking the logarithm inverse function, results in

$$X(t) = X(0) \cdot \exp\left(\alpha - \frac{\beta^2}{2}\right)t \cdot \exp(\beta W(t)), \tag{10.21}$$

in which $X(0)$ represents the initial value of the process, that remains to be found, based on the problem constraints.

The preceding formula can be written, in a simplified form, as

$$X(t) = X(0) \cdot e^{(\alpha - \frac{\beta^2}{2})t + \beta W(t)}. \tag{10.22}$$

This solution shows, for the proposed stochastic process, that the reject grows exponentially in time, controlled by the parameters α and β, that also remain to be determined. It is possible to note a random variation in the curve, as a result of the stochastic process $W(t)$.

For a stochastic process, $W(t)$, that corresponds to the total production line loss, if its distribution is Gaussian, it is possible to compute the associated reject distribution, using the transformation of the probability density function mathematical operation.

The expected value of the process, $X(t)$, is given by,

$$\begin{aligned} E[X(t)] &= E[X(0)] \cdot \exp\left(\alpha - \frac{\beta^2}{2}\right)t \cdot E[\exp(\beta W(t))] \\ &= E[X(0)]e^{\alpha t}. \end{aligned} \tag{10.23}$$

This last passage is obtained from the application of Itô's integration formula, as follows (Oksendal, 2013).

Initially, consider that

$$Y(t) = e^{\beta W(t)}, \tag{10.24}$$

and apply Itô's formula, to obtain

$$dY(t) = \beta e^{\beta W(t)} dW(t) + \frac{\beta^2}{2} e^{\beta W(t)} dt,$$

or,

$$Y(t) = Y_0 + \beta \int_0^t e^{\beta W(s)} dW(s) + \frac{\beta^2}{2} \int_0^t e^{\beta W(s)} ds.$$

But, by Theorem 3.2.1 (Oksendal, 2013),

$$E\left[\int_0^t e^{\beta W(s)} dW(s)\right] = 0, \tag{10.25}$$

therefore,

$$E\left[Y(t)\right] = E\left[Y_0\right] + \frac{\beta^2}{2}\int_0^t E\left[W(s)\right] ds,$$

or,

$$\frac{d}{dt}E\left[Y(t)\right] = \frac{\beta^2}{2}E\left[W(t)\right].$$

Finally,

$$E\left[Y(t)\right] = e^{\frac{\beta^2}{2}W(t)}. \tag{10.26}$$

Substituting the result 10.26 in the first line of Equation 10.23, one obtains the indicated result.

Therefore, according to the obtained mathematical model, the average value of the reject grows exponentially, for the specified conditions. The initial mean value of the process, $E[X(0)]$, should be determined from the data delivered by the industrial plant, as well as the estimate of the parameter, α, that controls the curve.

Figure 10.8 illustrates a stochastic process, that represents the evolution of the reject with time, as well as, the initial mean value of the process $E[X(t)]$, as a dotted curve.

Note that the stochastic process is non-stationary, in the long run, because its average value depends on time. Besides, the probability distribution of process $X(t)$, which represents the loss, is not Gaussian, as is usually assumed in the computation of losses using the usual theory of process statistical control.

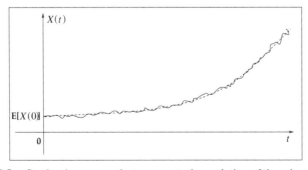

Figure 10.8   Stochastic process, that represents the evolution of the reject with time.

10.5.7 Diagnostic

From the preliminary analysis of the temporal series, one can verify that it is not possible to use the classic linear prediction method to model the reject in a production line, because the stochastic process is non-stationary.

Of course, the linear prediction model can be used to forecast in the short run, because it does not rely on production parameters. But, as mentioned, it requires that the process that models que reject be stationary in the considered interval.

The proposed model, which uses the Itô stochastic integral, allows to obtain a mathematical formula that reveals an exponential growth for the average value of the reject, that is controlled by production parameters, which can be determined from statistical moments that depend on the nature of the productive process. The method can be used to estimate the reject generation in other type of industry, with different products and distinct production lines.

The stochastic modeling of the production process permits to predict the short, or medium, term behavior of the reject series. It can also estimate unusual situations, periodicities, trends, and changes that are not visible in the observation of the raw series. This can be used to reduce the losses in the production line to a limited percentage of the total production, and built and automated control procedure, as compared to the manual evaluation and correction procedure currently used.

10.5.8 Obtained Results

The model that uses stochastic integration permits a medium to long term prediction of the reject series, which represents an advantage in relation to the traditional statistical methods, such as the process statistical control, but requires the determination of specific parameters, which depend on data provided by the company.

In the studied case, the long-term growth of the reject follows an exponential curve, that presents a stochastic variation with time. That variation depends on the addition of all the errors along the production process.

Figure 10.9 shows the measured evolution of the reject, with time, for the *Traditional* type of sandal, the best seller. It can be seen that the average loss follows an exponential curve. Figure 10.10 shows the measured evolution of the reject, with time, for the *Color* type of sandal. As the curve shows, the average loss also follows an exponential curve, with a different pattern. Figure 10.11 shows the measured evolution of the reject, with time, for the

Casual type of sandal. The average loss follows an exponential curve, with a high value for the α parameter.

The mathematical model indicates the corrective measures that should be taken to adjust the production line, and reduce the reject. Figure 10.12 shows the measured evolution of the reject, with time, for some sandals, after the correction measures are in place.

Although the instantaneous variation presents a Gaussian characteristic, because it is the result of multiple effects along the productive chain, the final process, which models the reject, does not have a Gaussian distribution. Its distribution is likely to be asymmetric.

The modeling indicates that it could be necessary to define an adequate percentage of reduction in the recycled material that is re-inserted in the

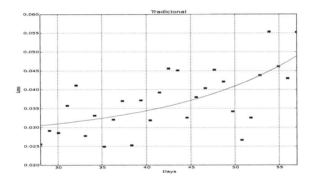

Figure 10.9   Measured evolution of the reject, with time, for the *traditional* type of sandal.

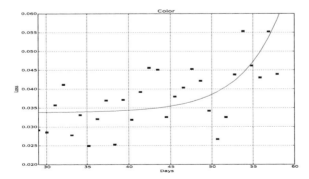

Figure 10.10   Measured evolution of the reject, with time, for the *color* type of sandal.

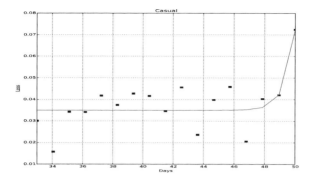

Figure 10.11 Measured evolution of the reject, with time, for the *casual* type of sandal.

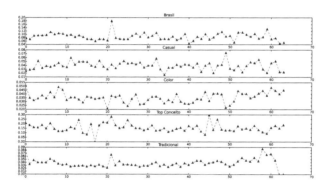

Figure 10.12 Measured evolution of the reject, with time, for a few sandals, after the correction measures are taken.

production line, such that, the reject is kept within specifications, for a given period of time.

Besides, from the detailed analysis and processing of the series, it is possible to indicate the most adequate periods to adjust the recycling process, using the reject.

10.6 Evaluation of the Stochastic Models

Regarding the preliminary analysis of time series, one can verify that the two modeling possibilities can be useful, and other models could, of course, be used, such as the crossing rate at the intended level of the loss (reject) curve.

The first model, which uses linear prediction, is intended for short term forecast and does not depend on the production parameter settings. But, it requires that the process which models the losses be stationary in the considered interval.

The second model, that uses Itô's stochastic integration, permits to obtain a model for the exponential growth of the reject mean value, which is controlled by parameters that must be determined from the statistical measures, and depend on the nature of the productive process.

The stochastic modeling of the productive process permits to forecast the short or mean time behavior of the reject series, with the possibility to estimate unusual situations, periodicities, biases, changes that are not clear to the analyst when the series is observed.

The autocorrelation model provides a short-term prediction for the losses, based only on the production time series. It is a linear prediction, based on the minimization of the mean square error. This procedure is optimum for the established conditions.

The stochastic integration model allows a mean to long term prediction for the losses, that is an advantage regarding the traditional methods of analysis, such as statistical control of processes, but needs to determine some specific parameters, which depend on measured data.

In the studied case, the long-term growth of losses follows an exponential curve, that includes a stochastic variation along time. This variation depends on the sum of errors in the production process.

Although the variation has a Gaussian characteristic because it is a result of multiple effects along the productive chain, the final process, that models the losses has an asymmetric distribution.

The modeling indicates that it must be necessary to define an adequate percentage of recycled material to be fed back in the production process, in order for the losses to be kept inside the technical specifications for a certain period of time.

Besides, from a detailed analysis of the series, it is possible to indicate the mode adequate periods to make the adjustments in the recycling process using the production losses.

10.7 The Influence of the Combination of Errors

As explained, the stochastic process $W(t)$ results from the combination of the errors, $W_i(t)$, which appear during the production process

$$W(t) = \sum_{i=1}^{N} W_i(t), \tag{10.27}$$

and has a Gaussian probability distribution. Therefore, Formula 10.22 can be written as

$$X(t) = X(0) \cdot e^{(\alpha - \frac{\beta^2}{2})t} \cdot e^{\beta W(t)}$$

$$= X(0) \cdot e^{(\alpha - \frac{\beta^2}{2})t} \cdot Y(t). \tag{10.28}$$

This explains how the combination of errors influence the reject final distribution, considering that the first term in the equation is a constant, the second term is a function of time, but deterministic, and the third term, which is also a function of time, is given by

$$Y(t) = e^{\beta W(t)} = \exp\left[\sum_{i=1}^{N} \beta W_i(t)\right] = \prod_{i=1}^{N} e^{\beta W_i(t)}, \tag{10.29}$$

an exponential function of the combination of errors $W(t)$, controlled by parameter β.

In other words, if adequate measures are taken to compensate for the exponential growth of the reject along time, by adjusting the parameters to obtain $\alpha = \frac{\beta^2}{2}$, the remaining term is $Y(t)$, which refers to the sum of errors in the production line, that has a Normal distribution.

Because $W(t)$ is a Gaussian stationary stochastic process, it is possible to obtain the distribution of the random process $Y(t)$, resulting from the transformation of the original distribution of combined errors.

Consider that process $W(t)$ has a probability density function $p_W(w)$, given by

$$p_W(w) = \frac{1}{\sigma_W \sqrt{2\pi}} e^{-\frac{(w - m_W)^2}{2\sigma_W^2}}, \tag{10.30}$$

in which the mean value of the random process $W(t)$ is given by $m_W = \mathrm{E}[W(t)]$, and the variance is $\sigma_W^2 = \mathrm{E}[(W(t) - m_W)^2]$.

Because the resulting process $Y(t)$ has distribution $p_Y(y)$, then, it is possible to use the property of probability density function transformation, to obtain

$$p_Y(y) = \frac{p_W(w)}{|dy/dw|}, \quad w = f^{-1}(y). \tag{10.31}$$

The derivative of the output process in relation to the input one, in simplified notation, is given by

$$\frac{dy}{dw} = \beta e^{\beta w}.$$

After a substitution, one obtains

$$p_Y(y) = \frac{e^{-\frac{(w-m_W)^2}{2\sigma_W^2}}}{\sigma_W\sqrt{2\pi}|\beta e^{\beta w}|}, \quad w = \frac{\ln y}{\beta}.$$

Substituting the inverse function, considering that β is positive, and that the exponential function is positive definite, leads to

$$p_Y(y) = \frac{e^{-\frac{(\frac{\ln y}{\beta}-m_W)^2}{2\sigma_W^2}}}{y\sigma_W\sqrt{2\pi}},$$

after some simplification, results in

$$p_Y(y) = \frac{e^{-\frac{(\ln y - \beta m_W)^2}{2\beta^2\sigma_W^2}}}{y\sigma_W\sqrt{2\pi}}, \quad \text{se } y > 0, \text{ e null case } y \le 0. \tag{10.32}$$

which represents the Lognormal probability density function.

The mean value of the random process $Y(t)$, for the resulting distribution, is given by

$$m_Y = \mathrm{E}[Y(t)] = \exp\left[\beta m_W + \frac{\beta\sigma_W}{2}\right]. \tag{10.33}$$

Note that the mean value of process $Y(t)$ grows with an exponential of the combination of the mean and standard deviation, adjusted by the parameter β, in relation to the original process $W(t)$, which represents the combination of effects of the production errors.

The variance of the random process $Y(t)$ is given by

$$\begin{aligned}
\sigma_Y^2 &= \mathrm{E}[(Y(t) - m_Y)^2] \\
&= \exp\left[2\beta m_W + \beta^2\sigma_W^2\right]\left[\exp(\beta^2\sigma_W^2) - 1\right]. \tag{10.34}
\end{aligned}$$

It is possible to observe that the combination of errors in the production line is a function of the product of the exponential of the sum of the original mean and variance adjusted, by the exponential of the variance minus one.

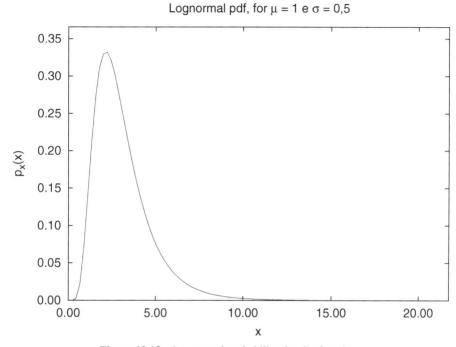

Figure 10.13 Lognormal probability density function.

Figures 10.13 and 10.14 illustrate, respectively, the Lognormal probability density function (pdf) and its cumulative probability function (CPF), in which $\mu = \beta m_W$ e $\sigma = \beta \sigma_W$.

In other words, the random process that represents the reject in a production line has a Lognormal distribution, instead of the Gaussian assumption for the reject.

This mistake in the modeling of the production line is usually responsible for problems associated with excessive product losses, that can compromise up to 12% of the company financial results.

10.8 Model to Estimate the Number, Lifespan or Creation Rate of New Businesses

Economic parameters, such as the number of companies in a given year, rate of opening of new businesses, or number of bankruptcies, are characterized as non-stationary stochastic processes. This modeling is usual for developing

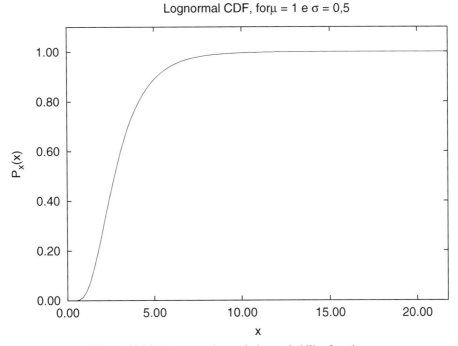

Figure 10.14   Lognormal cumulative probability function.

countries, in which the population did not reach stability. Developed countries are typically in statistical equilibrium. There is a strong correlation between the rate of opening of new businesses and economic growth. The growth increases the demand for products, which induces the opening of new companies to produce them. The creation of businesses induces the growth of the economy because it generates more jobs and revenue. This section presents a model to computate economic parameters, such as the number of active companies. The model can be used to predict the number of companies, or the bankrupt rate in the future or to estimate some economic indices.

10.8.1 Introduction

In the analysis of the evolution of institutions, one notices that it is not possible, theoretically, to depart from an initial state, free of institutions, and reach the state when the institutions finally emerge. It is proposed that, in the real world, the emergence of institutions is aided by concurring habits, as a result of emerging restrictions (Hodgson, 2001).

The level of economic activity in a market economy, according to John Maynard Keynes, depends on the willingness of entrepreneurs to invest. Economic problems occur when the interest is blocked by any difficulty to accumulate capital to make the investments (Heilbroner and Thurow, 1982) It is important to predict those problems, in due time, to provide alternatives or stimuli to the economic activity.

The detection of problems depends on the development of adequate mathematical models that allow forecasting the behavior of stochastic variables, such as the new businesses creation rate in a certain period or the number of businesses in a given year. But, there are few published results in the literature on the use of probability and stochastic processes to predict those economic variables (Alencar, 2009).

Models from queuing theory are used to study the difference between the growth rate of different countries, as well as, to evaluate the changes that occur with time. Markov chains, in which the transition probability matrix contains information from the economy, are helpful to such intent (Norris, 1997). Time-varying matrix parameters have been estimated, conditioned on the level of investment and quality of institutions (Morier and Teles, 2011).

An econometric study showed that the tendency to innovate grows with the size of the companies, for a set of companies that interact with universities and research institutes, according to a hypothesis attributed to Joseph Schumpeter. That is, large companies are more innovative than micro and small companies, regarding product innovation. But, this advantage disappears when dealing with process innovation (Póvoa and Monsueto, 2011).

Large companies have, on average, a longer lifespan than small ones. This can be the result of cooperation with universities and research centers, and also larger investment in innovation, that generates better products and higher revenue. The study of the growth in number and strength of the companies may direct the government actions, to elaborate public policies for the sector.

The decision to open a business depends on opposite forces: the prospect to obtain high revenues, which is an incentive, and the fear of failure, for those who avoid the risk. The market equilibrium usually requires a small number of businesses in the intermediary sector. This has an implication on the long-term growth of the economy, which is a function of the portion of entrepreneurs in the population.

The risk of self-employment is also shown in the high failure rates of the new ventures. According to the Panel Study of Income Dynamics (PSID), published in the United States, the failure rate in the first year, in 2000, was 35%. The entrepreneur's income is more volatile than the paid employee

(Clemens and Heinemann, 2006). Some studies indicate that an affluent economy may fail to produce the right incentives to the low-income economic agents, considering demand and supply (Chowdhury, 2013).

A study of business growth cycles in the USA, using a qualitative multivariate Hidden Markov Model (HMM), concluded that the markets usually detect economic inflection points, with a minimum period of three to six months, which is an incentive to adopt stochastic models (Bellone and Gautier, 2004).

Despite the direct relation between the microeconomics factors, such as, the number and dynamics of companies, and the macroeconomic factors, such as, the GNP, there are few studies that adopt probabilistic models to characterize the number, growth rate, and lifespan of the companies.

This section presents a model to predict the number of establishments in an economy based on the usual parameters of the economy. It discusses the methodology to develop the model, presents the mathematical model, introduces some results, and confirms real statistical data.

10.8.2 Methodology

The methodology to develop the mathematical model is based on queuing theory, which is a branch of probability that study the formation of queues or the storage of products, for example (Alencar, 2009).

The theory provides the means to predict the behavior of a system that offers, for instance, services whose demand grows randomly. This allows the design and dimensioning of the system to satisfy the clients and also to be profitable for the system provider, avoiding waste and bottlenecks.

The difference between the number of open and closed businesses, in a given period of time is the number of companies that remain economically viable, and integrate the total stock of the country.

It is possible to forecast the dynamics of the process using queuing theory, based on an adequate economic model. As a result of the mathematical modeling, it is possible to obtain the probability distribution of the states of the system, in this case, the number of establishments.

10.8.3 The Stochastic Model

The analysis of the behavior of establishments, from birth to death, requires that some quantities be measured directly in a finite observation interval $[0, T]$. For a number of companies $A(t)$, that are opened in a period t, and for a number of companies $B(t)$, that are closed in the same interval, the number of remaining companies in the economy is

$$R(t) = A(t) - B(t). \tag{10.35}$$

The quantities are stochastic variables, and the statistical average is given by

$$R = \mathrm{E}[R(t)] = \mathrm{E}[A(t)] - \mathrm{E}[B(t)] = A - B, \tag{10.36}$$

in which R is the mean number of remaining companies, for an average of A new companies, and an average of closed companies B, in the period $[0, T]$.

Therefore, the total lifespan of the companies in an economy is computed as the integral of the number of companies in operation in the time interval $[0, T]$,

$$\Gamma = \int_0^T R(t)dt = \int_0^T [A(t) - B(t)]dt. \tag{10.37}$$

The quantities are allowed to change from an observation period to another, but the relations are still valid. The average creation rate of new businesses (birth) in the interval $[0, T]$ is

$$\alpha = \frac{A}{T} \tag{10.38}$$

The average company lifespan is the ratio between the total time, and the birth rate of businesses

$$T = \frac{\Gamma}{A}, \tag{10.39}$$

Taking into account the previous definitions, the number of companies in a given economy, in the interval $[0, T]$, is then

$$N = \frac{\Gamma}{T} = \frac{\Gamma}{A} \cdot \frac{A}{T} = T \cdot \alpha. \tag{10.40}$$

That is, for an economic system observed in an interval $[0, T]$, which is empty at the origin and at time T, for $0 < T < \infty$, Formula No. (10.40) applies.

Note that N is a time average of the number of companies in the system, which is dimensionless, T is a sample mean of the lifespans of all the companies, measured in time units, and α is an average rate obtained from the count of companies that enter the market in the interval $[0, T]$, divided by T, the birth rate, and has a dimension of companies per unit time.

Formula 10.40, which allows the computation of the number of companies in an economic system in equilibrium, was developed by John Dutton Conant Little, to analyze the flow of packets in computer networks. The formula can be used to relate the number of companies in a certain economy with the average lifespan of the companies. In other words, the average number of companies equals the rate of creation of companies multiplied by the company mean lifespan (Little, 2011).

The formula does not require stationarity and is independent of the discipline of market formation by the companies. The company average lifespan equals the company mean operation time plus the time to open the company.

10.8.4 Results

According to the Statistics of U.S. Businesses (SUSB), published by the United States Census Bureau, the United States had a total of 5,684,424 firms and a total of 7,354,043 establishments, in 2011, which enrolled 113,425,965 employees, and produced a total payroll of U$ 5,164,897,905.00.

The 2011 annual or static data include the number of firms, number of establishments, employment, and annual payroll for most U.S. business establishments. The data are tabulated by geographic area, industry, and enterprise employment size. Industry classification is based on 2007 North American Industry Classification System (NAICS) codes (Bureau, 2011).

U.S. businesses numbered 7.4 million, in 2010, down by 36,800 from the previous year. The decline between 2008 and 2009 was 168,000 establishments, according to the Bureau's "County Business Patterns: 2010" report. Over 630,700 new businesses were started, during the fiscal year of 2010, in the United States of America, an increase from 585.500 in the previous year, according to data from The U.S. Small Businesses Administration (SBA) (SBA, 2014).

Total employment was 112 million, in 2010, a decline of 2.5 million workers. The drop from 2008 to 2009, at the height of the U.S. recession, was 6.4 million employees.

About three-quarters of all U.S. business firms have no payroll. Most are self-employed persons operating unincorporated businesses, and may or may not be the owner's principal source of income. Because non-employers account for only about 3.4% of business receipts, they are not included in most business statistics, for example, most reports from the Economic Census.

The average lifespan of a company listed in the S&P 500 index of leading US companies has decreased by more than 50 years in the last century, from 67 years in the 1920s to just 15 years today, according to Professor Richard Foster from Yale University

Introducing, into Formula 10.40, the figures for the number of establishments and also the number of establishments less than one year old, for the fiscal year of 2009, one obtains an estimate for the average lifespan of establishments in the United States

$$N = T \cdot \alpha,$$

or

$$T = \frac{N}{\alpha} = 7.4/0.5855 = 12.64.$$

That is, the average lifespan of the establishments in the United States, in 2009, was 12.64 years.

The formula is also useful to predict the behavior of the economy. Using data from 2009, it is possible to predict the number of companies in 2010, as follows.

$$T \cdot \alpha = 12.64 \times 630700 = 7972048.$$

which gives a small prediction error, when compared with the real figure.

10.8.5 Discussion

The number of companies in a given year, as well as the creation rate of new businesses can be characterized as stationary stochastic processes for developed countries, when the economy is in statistical equilibrium, or stationary.

The number of companies depends on several factors, related to the economy dynamics, such as the population consumption level, the employment rate, the inflation rate, the Gross National Product (GNP) growth, the emergence of new technologies, the aging of the economy, aside from other issues. Therefore, it is important to apply new probabilistic methods, mainly those that are simple, powerful, and easy to implement.

There is a strong correlation between the rate of opening of new businesses and economic growth. The growth increases the demand for products, which induces the opening of new companies to produce them. The creation of companies improves the economy because it generates more jobs and revenue.

This section presented a model to compute economic parameters, such as the number of active companies. The model can be used to predict the number of establishments, the bankrupt rate in the future. It can also be used to estimate indices that were not computed for a given year. For instance, if the number of companies created in a certain year is not known, it is possible to use the businesses lifespan, and their total number, to do the computation.

A

Probability Theory

"I cannot but regard determinism as a modern superstition."
Sir Karl Popper

A.1 Basic Probability Theory

Girolamo Cardano (1501–1576) wrote the first known book on probability, entitled *De Ludo Aleae* (About Games of Chance). He was an Italian medical doctor and mathematician, and the book was published, in 1663, nine decades after his death. Cardano was a known card player, and the manuscript was a handbook on this subject, containing some discussion on probability.

The first mathematical treatise on probability theory, published in 1657, was written by the Dutch scientist Christian Huygens (1629–1695), a manuscript entitled *De Ratiociniis in Ludo Aleae* (About Reasoning in Games of Chance).

Abraham de Moivre (1667–1754) was an important mathematician who worked on the development of Probability Theory. He wrote a book of great influence in his time, called *Doctrine of Chances*. The law of large numbers was discussed by Jacques Bernoulli (1654–1705) in his work Ars Conjectandi (The Art of Conjecturing).

Bernoulli, a Swiss mathematician of a family of prominent mathematicians, worked on the development of infinitesimal calculus, applying it to new problems. He published the first solution of a differential equation, that opened the path to the calculation of the variations of Leonhard Paul Euler (1707–1783) and Joseph Louis Lagrange (1736–1813), and extended their main applications to probability calculations (Dunham, 1990).

The study of probability was improved in the centuries XVIII and XIX, mainly because of the works of French mathematicians Pierre-Simon de Laplace (1749–1827) and Siméon Poisson (1781–1840), as well as the German mathematician Karl Friedrich Gauss (1777–1855).

A.2 The Axioms of Probability

The basic axioms of probability were established by Andrei Nikolaevich Kolmogorov (1903–1987), and allowed the development of the complete theory. Kolmogorov's first paper on probability appeared, in 1925, it was published jointly with Aleksandr Yakovlevich Khinchin (1894–1959), and contains the results on inequalities of partial sums of random variables which became the basis for martingale inequalities and the stochastic calculus.

Khinchin published several papers on measure theory and, in 1927, collected his contributions to this area in the paper *Recherches sur la Structure des Fonctions Mesurables*, published in the journal Fundamenta Mathematicae, of the Polish Academy of Sciences. This paper came out the same year he was appointed as a professor at Moscow University. In the same year, he also developed the basic laws of probability theory. In 1934, Khinchin published the foundations for the theory of stationary random processes in the journal Mathematische Annalen.

Kolmogorov was appointed a professor at Moscow University, in 1931. His thesis on probability theory, better known in its German translation from Russian, *Grundbegriffe der Wahrscheinlichkeitsrechnung* (Foundations of the Theory of Probability), published in 1933, put probability theory in a rigorous way, based on fundamental axioms.

In this work Kolmogorov managed to combine Advanced Set Theory, developed by Cantor, with Measure Theory, established by Lebesgue, to produce what is known as the modern approach to Probability Theory. The three fundamental statements are as follows (Papoulis, 1983):

1. Axiom 1: $P(\Omega) = 1$, in which Ω denotes the sample space or universal set and $P(\cdot)$ denotes the associated probability;
2. Axiom 2: $P(A) \geq 0$, in which A denotes an event belonging to the sample space;
3. Axiom 3: $P(A \cup B) = P(A) + P(B)$, in which A and B are mutually exclusive events and $A \cup B$ denotes the union of events A and B.

Kolmogorov established a firm mathematical basis on which other theories rely, including the Theory of Stochastic Processes, the Communications Theory, the Complexity Theory and the Information Theory, that use his axiomatic approach to Probability Theory.

The application of the axioms makes it possible to deduce all results relative to Probability Theory. For example, the probability of the empty set,

$\emptyset = \{\}$, can be calculated as follows. First, notice that

$$\emptyset \cup \Omega = \Omega,$$

since the sets $\emptyset$ and Ω are disjoint. Thus it follows that

$$P(\emptyset \cup \Omega) = P(\Omega) = P(\emptyset) + P(\Omega) = 1 \Rightarrow P(\emptyset) = 0.$$

In the case of sets A and B which are not disjoint it follows that

$$P(A \cup B) = P(A) + P(B) - P(A \cap B).$$

A.3 Bayes' Theorem

Thomas Bayes (1701–1761) was an English statistician, philosopher and Presbyterian minister, who formulated a special case of the theorem that bears his name. Bayes never published the theorem, but his notes were edited and published, in "An Essay Towards Solving a Problem in the Doctrine of Chances", which was read to the Royal Society of London, in 1763, after his death, by Richard Price (1723–1791), a Welsh moral philosopher and nonconformist preacher (Bayes, 1763).

Bayes' theorem was further improved by Pierre-Simon, marquis de Laplace (1749–1827), and defines the probability of an event, based on conditions that are given about that event. Bayes' rule, which is essential for the development of Information Theory, is a method to calculate conditional probabilities. It can be expressed by the following definition,

$$P(A|B) = \frac{P(A \cap B)}{P(B)},$$

assuming $P(B) \neq 0$.

An equivalent manner of stating the same result is the following,

$$P(A \cap B) = P(A|B) \cdot P(B) \ , \ \ P(B) \neq 0.$$

Some important properties of sets are presented next, in which A and B denote events from a given sample space.

- If A is independent of B, then $P(A|B) = P(A)$. It then follows that $P(B|A) = P(B)$ and that B is independent of A.
- If $B \subset A$, then: $P(A|B) = 1$.
- If $A \subset B$, then: $P(A|B) = \frac{P(A)}{P(B)} \geq P(A)$.

- If A and B are independent events then $P(A \cap B) = P(A) \cdot P(B)$.
- If $P(A) = 0$ or $P(A) = 1$, then event A is independent of itself.
- If $P(B) = 0$, then $P(A|B)$ can assume any arbitrary value. Usually in this case one assumes $P(A|B) = P(A)$.
- If events A and B are disjoint, and non-empty, then they are dependent.

A partition is a possible splitting of the sample space into a family of subsets, in a manner that the subsets in this family are disjoint and their union defines the sample space. It follows that any set in the sample space can be expressed with the use of a partition of that sample space and, therefore, be written as a union of disjoint events.

The following property can be illustrated by means of the Venn diagram, as illustrated in Figure A.1.

$$B = B \cap \Omega = B \cap \cup_{i=1}^{M} A_i = \cup_{i=1}^{N} B \cap A_i.$$

It now follows that

$$P(B) = P(\cup_{i=1}^{N} B \cap A_i) = \sum_{i=1}^{N} P(B \cap A_i),$$

$$P(A_i|B) = \frac{P(A_i \cap B)}{P(B)} = \frac{P(B|A_i) \cdot P(A_i)}{\sum_{i=1}^{N} P(B \cap A_i)} = \frac{P(B|A_i) \cdot P(A_i)}{\sum_{i=1}^{N} P(B|A_i) \cdot P(A_i)}.$$

Bayes' theorem is useful in several areas of study. It can be used to test hypotheses, such as those established for the occupancy of blank spaces in

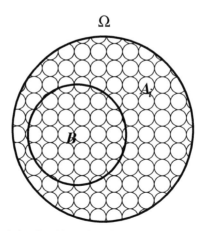

Ω

Figure A.1   Partition of set B by a family of sets $\{A_i\}$.

cognitive wireless sensor networks (Braga et al., 2015), or to evaluate the likelihood of certain diseases, for example (Joyce, 2008).

A.4 Random Variables

The idea of a random variable has the objective of transposing events of the sample space Ω to Borel segments, or rectangles, in the set of real numbers. This concept is defined informally, by many authors, so that readers can become acquainted with the idea before the appearance of formal applications in the theory.

In reality, what is called a random variable is not a variable, in the strict sense of the word, because it represents a function of random events, that are defined as a result of operations carried out in an algebra of sets. It could be more adequately defined as a dependent variable, the term independent variable is considered as representing the result of the experiment.

In addition, the variable is also not random, because it is a well-defined function, which is deterministic in this domain, and it usually has an inverse. However, the term random variable, although inaccurate, is still used in the literature.

The mapping $X : \Omega \mapsto \mathbb{R}$ is completely deterministic. This way, $X = f(\omega)$, in which $\omega \in \Omega$, as shown in Figure A.2. All the events of the sample space now have a counterpart in the set of real numbers, which is more adequate for usual operations, such as integration.

In the case of probability spaces, the expression "random variable" means a measurable function. That is, if $(\Omega, \mathcal{F}, P)$ is a probability space, then $X : \Omega \mapsto \mathbb{R}$ is a random variable if, for every $x \in \mathbb{R}$, the set $X^{-1}([x, \infty])$ is in $\mathcal{F}$, that is,

$$\{\omega \in \Omega : X(\omega) \geq x\} \in \mathcal{F}. \tag{A.1}$$

In the case in which $\Omega \subset \mathbb{R}$ is a measurable set, and $\mathcal{F} = \mathcal{B}$ is the σ-algebra of Borel subsets of Ω, the random variables are exactly the Borel functions $\mathbb{R} \mapsto \mathbb{R}$ (Capiński and Kopp, 2005).

Example: the experiment of rolling a dice and observing the result can be mapped into the set of natural numbers, $\mathbb{N}$, in the following manner

$$X(\omega_1) \to 1, X(\omega_2) \to 2, X(\omega_3) \to 3, X(\omega_4) \to 4, X(\omega_5) \to 5, X(\omega_6) \to 6.$$

because the function $X(\cdot)$ maps each outcome into a symbol of the set of natural numbers. ∎

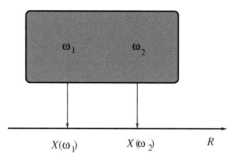

Figure A.2   Mapping of the events in the variable's domain.

This allows for mathematical operations and also makes it possible to plot graphs, so results can be analyzed. When analyzing the problem, it is important to distinguish between the elements of the sample space $\Omega = \{\text{heads}, \text{tails}\}$, which only indicate the sides of the coin and, therefore, it doesn't make sense to assign them probabilities, and the elements of the algebra, or family of subsets in the sample space, $\mathcal{F} = \{\emptyset, \Omega, \{\text{heads}\}, \{\text{tails}\}\}$, that represent the result of the experiment "tossing a coin and writing down the exposed side", to which is assigned the probability measure P. For example, $P(\{\text{heads}\}) = P\{\text{tails}\} = 1/2$.

Following what has been discussed, if f is a Borel function defined in the considered space and X is a random variable, then the function $f(X)$ is also a random variable (Rényi, 2007). To prove the statement, it suffices to consider, using a property of inverse functions, that $f^{-1}(X)(\mathcal{B}(\mathbb{R})) = X^{-1}\left[f^{-1}(X)(\mathcal{B}(\mathbb{R}))\right] \subset X^{-1}(\mathcal{B}(\mathbb{R})) \in \mathcal{F}$. Therefore, any function of a random variable is also a random variable.

A.5 Algebra Generated by a Random Variable

If X is a random variable, then $X^{-1}(\mathcal{B}) \subset \mathcal{F}$, but it can be a subset with a much smaller cardinality, depending on the complexity of X. This σ-algebra can be denoted as $\mathcal{X}$ and can be called the σ-algebra generated by X. The simplest case is that in which X is a constant, $X = c$, that produces a so called degenerate probability distribution. So, for a given set $B \in \mathcal{B}$, then $X^{-1}(\mathcal{B})$ is Ω or $\emptyset$, depending if $c \in B$ or not, and the σ-algebra generated is the trivial, $\mathcal{F} = \{\emptyset, \Omega\}$.

If X assumes different two values, a and b, then $\mathcal{X}$ contains four elements, $\mathcal{X} = \{\emptyset, \Omega, X^{-1}(\{a\}), X^{-1}(\{b\})\}$. If X can be one of many finite values, then $\mathfrak{X}$ is finite. If X assumes countable values, then $\mathfrak{X}$ is non-countable and can be identified with the σ-algebra of all the subsets of a countable set. One may notice that the cardinality of $\mathfrak{X}$ increases with the complexity of X.

It must be considered that the set of points in which leaps occur in a probability distribution is countable, which allows the attribution of a probability to all the points considered, and that the distribution function that increases only through leaps is called a discrete distribution. In addition, the distributions always have limits to the left, even if they do not coincide with the limits to the right of the discontinuity points.

Example: consider $X(\omega) = \omega$ and $Y(\omega) = 1 - \omega$, for $\Omega = [0, 1]$. Then, $\mathfrak{X} = \mathfrak{Y} = \mathfrak{F}$, and a σ-algebra cannot be independent from itself, apart from the trivial case.

Let $A \in \mathfrak{F}$, then the criterion of independence requires $P(A \cap A) = P(A) \times P(A)$, that is, the set A belongs to both σ-algebras. But this implies $P(A) = P(A)^2$, which can only happen if $P(A) = 1$ or $P(A) = 0$. Therefore, a σ-algebra independent from itself consists of sets with measure one or zero. ∎

It is important to mention that, although the distribution is the most important concept to characterize a random variable, it doesn't specify it completely. If, for example, the distributions of the random variables X and Y are known, it is not possible, in general, to determine the distribution of the variable $Z = f(X, Y)$ only from this information. For this, it is necessary to know the joint distribution of X and Y (Rényi, 2007).

A.6 Lebesgue Measure and Probability

The simple Lebesgue measure, previously discussed, can be used in the calculation of probabilities in the case of uniform distribution in the interval $U = [0, 1]$, defined in $\mathbb{R}$, because in this case, the probability measure is given by the length of the variable's interval exactly (Rosenthal, 2000). That is, $P(0 \leq X \leq 1/2) = 1/2$ and, in general for the uniform distribution, $P(a \leq X \leq b) = |b - a|, 0 \leq a \leq b \leq 1$.

Apart from this, for the disjoint subsets of U, $A = [a, b]$ and $B = [c, d]$, in which $0 \leq a \leq b \leq c \leq d \leq 1$, it is always true that $P(A \cup B) = |b - a| + |d - c|$, that is, finite additivity can be applied,

$$P(A \cup B) = P(A) + P(B). \tag{A.2}$$

Evidently, to allow for countable operations with sets, such as limits, which are important in probability theory, it is necessary to extend Equation A.2 to a countably infinite number of disjoint sets,

$$P(A_1 \cup A_2 \cup A_3 \cup \cdots) = P(A_1) + P(A_2) + P(A_3) + \cdots , \qquad \text{(A.3)}$$

which is called countable additivity. Uncountable additivity is not defined, because it would imply, for example,

$$P([0,1]) = \sum_{x \in [0,1]} P(\{x\}), \qquad \text{(A.4)}$$

which is clearly false, because the left-hand side of the equation is equal to 1, while the right-hand side is 0, seen that the measure of a point is zero (Rosenthal, 2000).

The following sections present a probability measure valid for any distribution, as long as the variable is defined in a Borel interval, or rectangle, in the set of real numbers. It is initially defined for an open subset of $\mathbb{R}$.

A.7 Cumulative Distribution Function

The Cumulative Distribution Function (CDF), represented in Figure A.3, is an application of the probability measure, in the probability space $(\mathbb{R}, \mathcal{B}(\mathbb{R}), P)$, to a specific set, that is,

$$P_X(x) = P\{x \in \mathbb{R} : -\infty < X \le x\} = P\{X \le x\}. \qquad \text{(A.5)}$$

Note that this definition is equivalent to that of a measurable function, induced by the measure in the probability space $(\Omega, \mathcal{F}, P)$, in which the reuse of P denotes a certain overindulgence of the language and the simplification $P(\{X \le x\}) = P\{X \le x\}$ is made (Marques, 2009),

$$
\begin{aligned}
P_X(x) &= P\{\omega \in \Omega : -\infty < X(\omega) \le x\} \\
&= P\{X^{-1}((-\infty, x]) \in \mathcal{F}, \forall x \in \mathbb{R}\}. \qquad \text{(A.6)}
\end{aligned}
$$

With the definition, a function in the set of real numbers can be obtained, instead of a mathematical operation between generic functions, which makes the process of measuring probability more operational and appropriate for calculating the means, among other operations.

The CDF has the following properties, which can be verified:

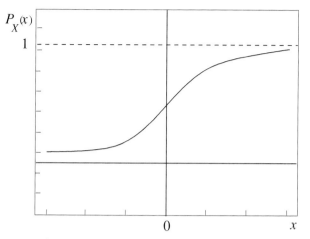

Figure A.3   Cumulative distribution function.

1. All the probabilities are exhausted in the universal set

$$P_X(\infty) = P\{-\infty < X \le \infty\} = P\{X < \infty\} = P(\Omega) = 1; \quad (A.7)$$

2. The probability of a point is zero,

$$P_X(-\infty) = P\{-\infty < X \le -\infty\} = P\{X < -\infty\} = P(\emptyset) = 0;$$
$$(A.8)$$

3. For the disjoint Borel segments, like

$$\{X > x\} \cup \{X \le x\} = \Omega,$$

that form a partition, then

$$P\{X > x\} + P\{X \le x\} = 1 \;\Rightarrow\; P\{X > x\} = 1 - P_X(x); \quad (A.9)$$

4. The probability that a random variable is in the interval $[a, b]$ is given by

$$P\{a < X \le b\} = P\{X \le b\} - P\{X \le a\} = P_X(b) - P_X(a).$$
$$(A.10)$$

The CDF $P_X(x)$ can be written as the primitive of a function, that is

$$P_X(x) = \int_{-\infty}^{x} p_X(\alpha)d\alpha.$$

Therefore, $P_X(a)$ represents the area under the curve up to point a. From the fundamental theorem of Calculus, it can be verified that

$$p_X(x) = \frac{d}{dx} P_X(x),$$

in which $p_X(x)$ is known as the probability density function (pdf), illustrated in Figure A.4. In this figure, the darkened area represents the calculation of the probability density function.

It is important to notice that the CDF provides a measure of probability; therefore, it is dimensionless. However, the pdf does have dimension, which is the inverse of the dimension of the random variable considered.

Example: the exponential probability density models packet arrival time in a computer network and, also, the duration of a call in a telephone system. This distribution is generally written as

$$p_X(x) = \alpha e^{-\alpha x} u(x),$$

in which $u(x)$ is the unitary step function, shown in Figure A.5.

An exponential random variable X, with parameter α, has $P\{X \geq x\} = e^{-\alpha x}$, which means that the probability that the variable exceeds a given value x is also exponential. This distribution has the property of being memoryless, that is, $P\{X > t + s | X > t\} = P\{X > s\}$. ■

Next, some important relations between the CDF and the pdf are presented:

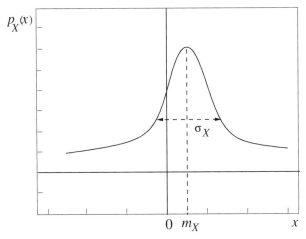

Figure A.4   The probability density function.

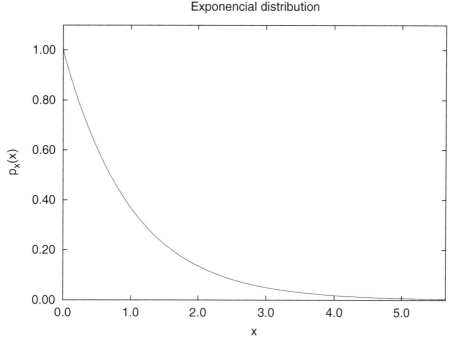

Figure A.5 Exponential probability density function, with parameter $a = 1$.

$$P\{X > x\} = 1 - \int_{-\infty}^{x} p_X(\alpha)\, d\alpha = \int_{-\infty}^{\infty} p_X(\alpha)\, d\alpha - \int_{-\infty}^{x} p_X(\alpha)\, d\alpha,$$

or

$$P\{X > x\} = \int_{x}^{\infty} p_X(\alpha)\, d\alpha.$$

Also,

$$P\{a < X \leq b\} = P_X(b) - P_X(a) = \int_{-\infty}^{b} p_X(x)\, dx - \int_{-\infty}^{a} p_X(x)\, dx,$$

which gives,

$$P\{a < X \leq b\} = \int_{a}^{b} p_X(x)\, dx.$$

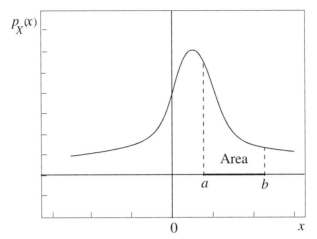

Figure A.6   The area under the probability density function, from a to b is the difference between the respective cumulative distributions.

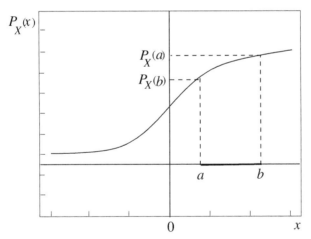

Figure A.7   The probability measurement using the cumulative distribution function.

Example: a voice signal can be modeled using a Laplace distribution, as follows, for which the previous property can be verified,

$$p_V(v) = \frac{a}{2}e^{-a|v|}.\blacksquare$$

The Laplace distribution was named after the French mathematician, astronomer and physicist Pierre-Simon, Marquis de Laplace (1749-1827),

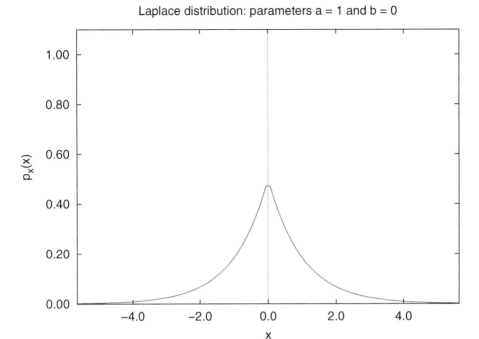

Figure A.8 Laplace probability density function, with parameters $a = 1$ and $b = 0$.

In the generic case, the Laplace distribution has two control parameters, to adjust the mean and the standard deviation of the distribution.

Figure A.8 illustrates the pdf for $a = 1$ and $b = 0$.

$$p_X(x) = \frac{a}{2}e^{-a|x-b|}. \tag{A.11}$$

For the Laplace distribution with zero mean, the probability that X is in the interval (c, d) is given by

$$P(c < x \le d) = \int_c^d p_X(x)\mathrm{d}x = \int_c^d \frac{a}{2}e^{-a|x|}\mathrm{d}x = 1 - \frac{1}{2}e^{-ac} + \frac{1}{2}e^{ad}. \tag{A.12}$$

The CDF is calculated as the integral of $p_X(x)$

$$P_X(x) = \int_{-\infty}^x p_X(x)\mathrm{d}x = \int_{-\infty}^x \frac{a}{2}e^{-a|x|}\mathrm{d}x = 1 - \frac{1}{2}e^{-ax}. \tag{A.13}$$

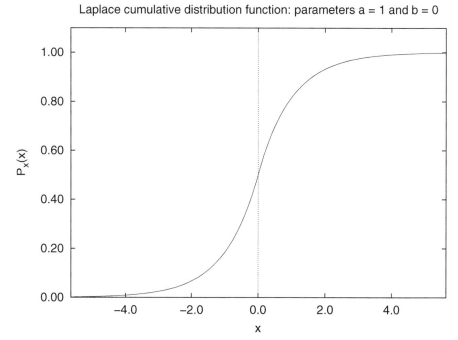

Figure A.9   Laplace cumulative distribution function, with parameters $a = 1$ and $b = 0$.

Figure A.9 illustrates the CDF for a Laplace distribution, considering the parameters $a = 1$ and $b = 0$.

A.7.1 Moments of a Random Variable

The moments of a random variable are statistics that reveal important information. They represent the variable's means, indicating trends and deviations. For example, considering the distribution of grades in a classroom, one can calculate the classroom's mean and also the deviation centered at this mean. In an election, it is possible to determine the voting trend. In a biological experiment, the confidence interval of the data can be calculated.

The mathematical expectation can be defined using the Lebesgue integral, particularized to the probabilistic measure space $(\Omega, \mathcal{F}, P)$. For such, some conditions of integrability are necessary for the random variable.

For the probability space $(\Omega, \mathcal{F}, P)$, consider the random variable $X :$ $\Omega \mapsto \mathbb{R}$, which takes the event $\{\omega\} \in \mathcal{F}$ to the random variable $X(\omega)$. The mathematical expectation, also known as the expected value, of X, is defined as

$$E[X] = \int_\Omega X dP = \int_\Omega X(\omega)dP(\omega). \tag{A.14}$$

The knowledge of the correspondence, that defines the random variable as a function of a random event, is not always necessary to determine its expected value. It is sufficient to know the law, or measure, of probability of the random variable.

Considering that the measure used is the CDF, that is, $\mu(x) = P_X(x)$, which is positive and additive, then

$$E[f(X)] = \int_\Omega f(X)dP = \int_\mathbb{R} f(x)dP_X(x). \tag{A.15}$$

A function $f(x) : \mathbb{R} \mapsto \mathbb{R}$ is integrable relative to the measure $\mu(x) = P_X(x)$, if and only if $f(X)$ is integrable relative to the measure P. The rightmost integral in Equation A.15 is, in some texts, called the Stieltjes integral, or Riemann–Stieltjes integral, or even the Lebesgue integral relative to the probability measure $P_X(x)$.

To demonstrate the validity of Equation A.15, a function $f = I_A$ can be considered, with $A \subset \mathcal{F}(\mathbb{R})$, and it can be verified that the equality results from having

$$f(X)(\omega) = I_A(X)(\omega) = \begin{cases} 1, & X(\omega) \in A, \text{ that is, if } \omega \in X^{-1}(A) \\ 0, & X(\omega) \notin A, \text{ that is, if } \omega \notin X^{-1}(A). \end{cases}$$

Therefore, $I_A(X) = I_{X^{-1}(A)}$, which implies

$$\int_\Omega f(X)dP = \int_\Omega I_{X^{-1}(A)}dP = P[X^{-1}(A)] = \mu(A) = \int_\mathbb{R} f(x)dP_X(x).$$

Let $f(X)$ be a random variable function, the expected value relative to X of the function $f(X)$, considering that the cumulative function $P_X(x)$ has the derivative $p_X(x)$, is given by

$$E[f(X)] = \int_{-\infty}^\infty f(x)p_X(x)dx. \tag{A.16}$$

A.7.2 Properties Associated to the Expected Value

As a result of the properties of the Lebesgue integral, the expected value operator, or mathematical expectation has the following properties

$$E[\alpha X] = \alpha \cdot E[X], \tag{A.17}$$

for a given constant α.

$$E[X + Y] = E[X] + E[Y], \tag{A.18}$$

for any given random variables X and Y, and

$$E[XY] = E[X] \cdot E[Y] \tag{A.19}$$

if X and Y are independent random variables. X and Y are considered independent if the σ-algebras generated by them are independent, that is, for any Borel sets B and C in $\mathbb{R}$,

$$P(X^{-1}(B) \cap P(Y^{-1}(C)) = P(X^{-1}(B)) \times P(Y^{-1}(C)). \tag{A.20}$$

Many moments of the random variable X, defined in the probability space $(\mathbb{R}, \mathcal{B}((\mathbb{R}), P)$ have special importance and distinct physical interpretations for different domains. In a general manner, the moments are defined by the formula

$$E[X^n] = \int_{-\infty}^{\infty} x^n p_X(x) \mathrm{d}x. \tag{A.21}$$

For a discrete probability distribution, of the type

$$p_X(x) = \sum_{k=-\infty}^{\infty} p_k \delta(x - x_k), \tag{A.22}$$

the expected value is calculated as

$$
\begin{aligned}
E[X^n] &= \int_{-\infty}^{\infty} x^n \sum_{k=-\infty}^{\infty} p_k \delta(x - x_k) \mathrm{d}x \\
&= \sum_{k=-\infty}^{\infty} \int_{-\infty}^{\infty} p_k x^n \delta(x - x_k) \mathrm{d}x \\
&= \sum_{k=-\infty}^{\infty} p_k x_k^n.
\end{aligned}
\tag{A.23}
$$

Note that the sampling property of the impulse function was used in the last step.

A.7.3 Definition of the Most Important Moments

The first moments are more well-known, usually being related to the means and physical measures. They are defined as:

- $m_1 = m_X = E[X]$, arithmetic mean, mean value, average voltage value of a signal X, statistical average.
- $m_2 = P_X = E[X^2]$, square mean, total power of a signal X.
- $m_3 = E[X^3]$, moment or measure of the asymmetry (skewness) of the probability density function.
- $m_4 = E[X^4]$, moment or measure of the flatness of pdf.

The variance, also known as the AC power of X in engineering, is an important statistical measure in Medicine, Engineering, Economics, and other fields. It is defined as follows,

$$V[X] = \sigma_X^2 = E[(X - m_1)^2] = m_2 - m_1^2, \tag{A.24}$$

in which the standard deviation, σ_X, is defined as the square root of the variance and indicates how much the variable deviates from its mean value.

The variance has the following properties, that can be verified using the definition,

$$V[\alpha X + \beta] = \alpha^2 V[X],$$

$$V\left[\sum_{i=1}^{N} X_i\right] = \sum_{i=1}^{N} V[X_i].$$

The coefficient of asymmetry, or skewness, is defined, for a random variable with expected value m_X and standard deviation σ, by using the corresponding moment (Magalhães, 2006)

$$\alpha_3 = \frac{E[(X - m_X)^3]}{\sigma_X^3}. \tag{A.25}$$

The kurtosis coefficient, which measures the intensity of the peaks in the probability distribution of a random variable X, is defined as

$$\alpha_4 = \frac{E[(X - m_X)^4]}{\sigma_X^4}. \tag{A.26}$$

The median of a random variable, m_D, is the measure of the position of the center of its distribution, and it divides the probability distribution into two parts with equal areas.

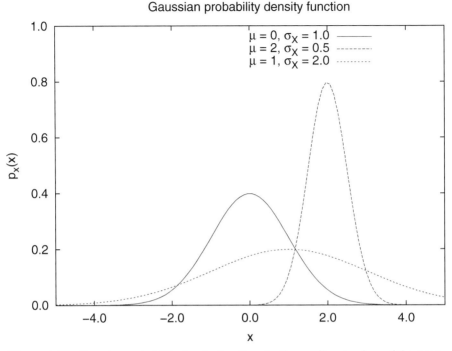

Figure A.10   Gaussian probability density function, for three different values of the mean, μ, and standard variation, σ_X.

$$\int_{-\infty}^{m_D} p_X(x)\mathrm{d}x = \frac{1}{2}. \tag{A.27}$$

This measure is frequently used in the calculation of populations, especially in sociology and economics. It is equal to the mean if the pdf is symmetric.

Example: consider the Gaussian, or Normal, distribution, illustrated in Figure A.10 for some values of the mean and standard deviation, the formula for which is given by

$$p_X(x) = \frac{1}{\sigma_X\sqrt{2\pi}}e^{-\frac{(x-m_X)^2}{2\sigma_X^2}}.$$

The mean value of the random variable is given by

$$\mathrm{E}[X] = \int_{-\infty}^{\infty} x p_X(x)\,\mathrm{d}x = \int_{-\infty}^{\infty} \frac{x}{\sigma_X\sqrt{2\pi}}e^{-\frac{(x-m_X)^2}{2\sigma_X^2}}\,\mathrm{d}x.$$

However, it is known that the area under the curve of the normal function is unitary,

$$\int_{-\infty}^{\infty} \frac{1}{\sigma_X \sqrt{2\pi}} e^{-\frac{x^2}{2\sigma_X^2}} \, dx = 1.$$

Therefore, with a change of variables, it can be seen that

$$
\begin{aligned}
E[X] &= \int_{-\infty}^{\infty} (y + m_X) \cdot \frac{1}{\sigma_X \sqrt{2\pi}} e^{-\frac{y^2}{2\sigma_X^2}} \, dy \\
&= \frac{1}{\sigma_X \sqrt{2\pi}} \left[\int_{-\infty}^{\infty} y e^{-\frac{y^2}{2\sigma_X^2}} \, dy + m_X \int_{-\infty}^{\infty} e^{-\frac{y^2}{2\sigma_X^2}} \, dy \right]. \quad (A.28)
\end{aligned}
$$

For the rightmost term, the first function to be integrated is odd in the considered interval, and the result of the integration is zero. The second integral is unitary because it represents the area under the pdf. Therefore,

$$E[X] = m_X.$$

Since the Gaussian distribution is symmetric in relation to the mean, the mean and the median are the same.

The CDF is calculated integrating the pdf,

$$P_X(x) = \frac{1}{\sigma_X \sqrt{2\pi}} \int_{-\infty}^{x} e^{-\frac{(\alpha - m_X)^2}{2\sigma_X^2}} \, d\alpha.$$

The result is shown in Figure A.11, for some mean and standard deviation values. ■

A.8 Functions of Random Variables

The theory of functions of a random variable also called the pdf transformation, is rich in applications in different areas of engineering. One application is illustrated in Figure A.12, which shows the operation executed by the circuit described by the equation $Y = g(X)$. The circuit can be a linear amplifier, an attenuator, a rectifier, a limiter, a logarithmic amplifier, or any other operation that does not involve energy storage devices, such as capacitors or inductors.

One method to calculate the probability distribution of the signal at the output of the circuit, assuming that it can be described by the associated

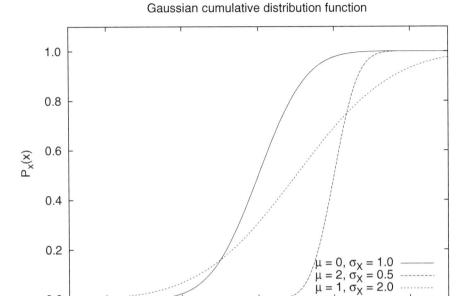

Figure A.11   Gaussian cumulative distribution function, for three different values of the mean, μ, and standard deviation, σ_X.

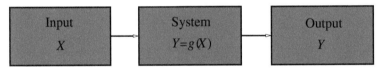

Figure A.12   Application of transformation of pdf to the solution of a circuit, described by the equation $Y = g(X)$.

random variable, is to use the cumulative function in terms of the measure of the Borel interval.

Consider an interval $(-\infty, y]$ and a probability measure P. It is known that the probability distribution $P_Y(y)$ of the output variable is the probability of the event $\{Y \leq y\}$, consisting of all the results ω such that $Y(\omega) = g[X(\omega)] \leq y$.

Therefore,

$$P_Y(y) = P\{Y \leq y\} = P\{g(X) \leq y\}. \tag{A.29}$$

There are a few necessary conditions so that $g(X)$ can be a random variable:

- The domain of $g(x)$ must include the image of the random variable X.
- The events $\{g(X) = \infty\}$ and $\{g(X) = -\infty\}$ must have zero probability.
- For $\{Y \leq y\}$ to be an event, the set of values of X for which $g(x) \leq y$ must be formed from the union and intersection of a countable number of Borel intervals.

If the requirements are met, and the function is bijective, the inverse $g^{-1}(y)$ can be applied in Equation A.29, to obtain

$$P_Y(y) = P\{X \leq g^{-1}(y)\} = P_X(g^{-1}(y)). \tag{A.30}$$

Example: a linear system is governed by the equation $Y = AX$. Given the distribution $P_X(x)$ for the input variable, what is the distribution for the output variable? What is the pdf of the output random variable?

From the Formula A.30, and assuming that the inverse function is given by $x = g^{-1}(y) = y/A$, then

$$P_Y(y) = P_X(y/A).$$

Differentiating the obtained result, and knowing that A can either be negative or positive, but not zero, yields

$$p_Y(y) = \frac{d}{dy}P_X(y/A) = \frac{p_X(y/A)}{|A|}.$$

For the case in which $A = 0$, which means there is no output signal, the probability measure is projected onto zero, and therefore the CDF is given by

$$P_Y(y) = u(y/A) = u(y),$$

and the pdf is

$$p_Y(y) = \delta(y). \blacksquare$$

The operation executed by the generic system shown in Figure A.12 can also be seen, in the domain of the random variable, as illustrated in Figure A.13. The probability distribution of the input variable is transformed by the system, generating a different distribution at the output.

Example: assuming that X is a random variable, with a CDF given by

$$P_X(x) = P(X \leq x) = \frac{1}{(1 + e^{-x})^{\alpha}},$$

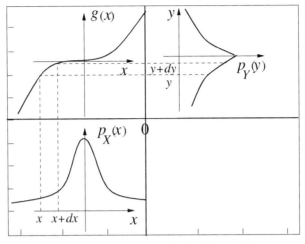

Figure A.13   The bijective mapping implies equivalent areas.

in which $\alpha > 0$ is a given parameter. Consider a system that produces the following random variable at the output

$$Y = \ln(1 + e^{-X}).$$

Then, by definition,

$$P_Y(y) = P(Y \le y) = P(\ln(1 + e^{-X}) \le y) = P(X > -\ln(e^y - 1)).$$

The inequality can be evaluated in terms of the cumulative distribution of X,

$$
\begin{aligned}
P_Y(y) &= 1 - P_X(-\ln(e^y - 1)) \\
&= 1 - \frac{1}{(1 + e^{\log(e^y - 1)})^\alpha} \\
&= 1 - \frac{1}{(1 + e^y - 1)^\alpha} \\
&= 1 - e^{-\alpha y}, \text{ for } y \ge 0.
\end{aligned}
\tag{A.31}
$$

The last formula is the CDF of an exponential distribution, and the pdf is obtained by differentiating $P_Y(y)$, which gives

$$p_Y(y) = \frac{dP_Y(y)}{dy} = e^{-\alpha y}, \text{ for } y \ge 0. \blacksquare$$

Since the bijective mapping implies that the areas shown are equivalent, then

$$p_X(x)\, dx = p_Y(y)\, dy,$$

for which the solution is shown next, taking into account that the inversion of the curve of the system requires a change in the sign of the derivative,

$$p_Y(y) = \frac{p_X(x)}{|dy/dx|}, \quad x = g^{-1}(y).$$

The formula presented assumes the existence of the inverse of $f(\cdot)$ as well as its derivative at all points. Because the probability is always positive and dy/dx can be negative, the absolute value of the derivative must be taken. On the other hand, it is necessary to obtain $x = g^{-1}(y)$, because an answer is desired in terms of Y and the input is in terms of X.

Example: a random variable X is amplified by a linear system with positive gain G, that is,

$$Y = GX.$$

Obtain the pdf of Y, knowing that

$$p_X(x) = \alpha e^{-\alpha x}\, u(x).$$

$$\frac{dy}{dx} = G$$

$$x = \frac{y}{G}$$

Using the formula for the transformation of the pdf, yields

$$p_Y(y) = \frac{\alpha e^{-\alpha \frac{y}{G}} u(y/G)}{G}.$$

Therefore,

$$p_Y(y) = \frac{\alpha}{G} e^{-\frac{\alpha y}{G}} u(y). \blacksquare$$

Example: consider an amplifier with a bias, B, modeled by the equation $Y = GX + B$, in which the gain, G, is positive. Note that the function is not linear, it is rather known as an affine function. A function is linear if it is:

1. homogeneous, that is, $\alpha x \to \alpha y$;
2. additive, that is, $x_1 + x_2 \to y_1 + y_2$.

The solution for the problem is obtained by following the steps illustrated in Figure A.14:

1. Obtain the differential gain, or differentiate the output in relation to the input. In this case, $\frac{dy}{dx} = G, \ G > 0$;
2. Obtain the inverse of the function that relates the input to the output, that is, $x = \frac{y-B}{G}$;
3. Apply the formula $p_Y(y) = \frac{p_X(x)}{|dy/dx|} = \frac{p_X((y-B)/G)}{G}$. ■

Example: substituting the Gaussian distribution from Figure A.10 in the previous example, in which $Y = GX + B$, results in

$$p_X(x) = \frac{1}{\sigma_X\sqrt{2\pi}}\,e^{-\frac{x^2}{2\sigma_X^2}}.$$

$$p_Y(y) = \frac{p_X(x)}{\left|\frac{dy}{dx}\right|},$$

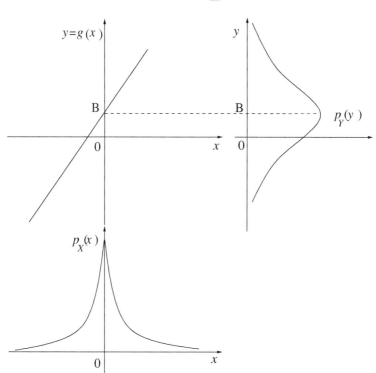

Figure A.14   Illustration for the pdf transformation caused by a linear amplifier.

$$y = Gx + B \Rightarrow x = \frac{y - B}{G}, \qquad \frac{dy}{dx} = G > 0.$$

Therefore,

$$p_Y(y) = \frac{p_X\left(\frac{y-B}{G}\right)}{G},$$

$$p_Y(y) = \frac{1}{\sigma_X \sqrt{2\pi}G} \, e^{-\frac{\left(\frac{y-B}{G}\right)^2}{2\sigma_X^2}},$$

Using $\sigma_X G = \sigma_Y$, yields

$$p_Y(y) = \frac{1}{\sigma_Y \sqrt{2\pi}} \, e^{-\frac{(y-B)^2}{2\sigma_Y^2}} . \blacksquare$$

Considering X as an electrical signal, the following relationships can be verified:

- The total power of the input signal is $P_X = \sigma_X^2$.
- The average power, also called direct current (DC) power, of the input signal is $P_{DC} = m_X^2 = 0$.
- The variance is the AC power, given by $P_{AC} = V[X] = \sigma_X^2$.

For the output signal:

- The variance is given by $P_{AC} = V[Y] = \sigma_Y^2 = G^2 \sigma_X^2$, which results in an effective value $V_{RMS} = \sigma_Y$.
- The mean level of the output signal is $P_{DC} = m_Y^2 = B^2$.
- The total power of the output signal is $P_Y = G^2 \sigma_X^2 + B^2$. It is a function of the square of the gain and the square of the bias voltage.

A.9 General Formula for Transformation

The general formula for the calculation of the pdf transformation, which includes the cases where the function has partial only inverses or does not have an inverse for the entire interval can be obtained with the use of the impulse function, created by Paul Adrien Maurice Dirac (1902–1984), an English electrical engineer and mathematician, who graduated from Bristol University.

It is important to take into account that the function has inverses only in the intervals i, and that they are given by $f_i^{-1}(y)$, while it is constant for the intervals j, and hence the function does not have an inverse in those intervals.

Therefore, the output presents impulses for the points y_j, with areas a_j. Then, the pdf of the system response is given by

$$p_Y(y) = \sum_i \frac{p_X(x_i)}{|dy/dx_i|} \Big|_{x_i=f_i^{-1}(y)} + \sum_j a_j \delta(y - y_j). \qquad \text{(A.32)}$$

The preceding formula permits the computation of the output probability function for any reasonable transformation $y = f(x)$. Of course, there are functions, such as Dedekind's, which require the use of the Lebesgue integral.

Example: if it is known that $Y = X^2$, and that $p_X(x) = \frac{1}{\sqrt{2\pi}} e^{\frac{-x^2}{2}}$, $-\infty < x < \infty$, determine $p_Y(y)$.

$$p_Y(y) = \frac{p_X(x)}{|dy/dx|} \Big|_{x=f^{-1}(y)} \cdot$$

$$|\frac{dy}{dx}| = |2x|.$$

Since the function $y = x^2$ does not admit an inverse, the problem must be solved in parts and the results added.

Region I: $x = \sqrt{y}$,

$$p_{Y1}(y) = \frac{1}{2\sqrt{2\pi y}} e^{-\frac{y}{2}}.$$

Region II: $\rightarrow x = -\sqrt{y}$,

$$p_{Y2}(y) = \frac{1}{2\sqrt{2\pi y}} e^{-\frac{y}{2}}.$$

Now,

$$p_Y(y) = p_{Y1}(y) + p_{Y2}(y),$$

which results in

$$p_Y(y) = \frac{1}{\sqrt{2\pi y}} e^{-\frac{y}{2}}, \ 0 < y < \infty. \blacksquare$$

This pdf is called the chi-square distribution and has the following general formula

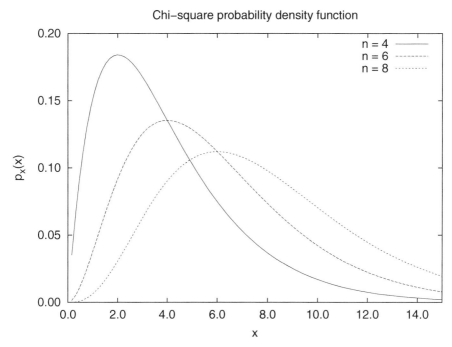

Figure A.15 Chi-square probability density function, for three different values of the parameter n.

$$P_X(x) = \frac{x^{n/2-1}e^{-x^2}}{2^{n/2}\Gamma(n/2)}, \ 0 < x < \infty, \tag{A.33}$$

in which

$$\Gamma(n) = \int_0^\infty x^{n-1}e^{-x}\mathrm{d}x, \ n > 0. \tag{A.34}$$

The distribution is shown in Figure A.15. It is typically used in hypothesis testing for biometrics to find, for example, the dispersion value of two nominal variables, evaluating the existing association between qualitative variables.

Two groups are said to behave in a similar manner if the difference between the frequencies observed and the frequencies expected in each category is small. The chi-square CDF is illustrated in Figure A.16.

Example: a signal with a Gaussian distribution is applied to a full-wave rectifier circuit.

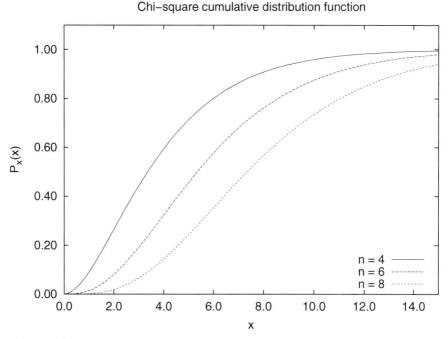

Figure A.16   Chi-square cumulative distribution function, for three different values of the parameter n.

Once again considering the signal has the distribution,

$$p_X(x) = \frac{1}{\sigma_X \sqrt{2\pi}}\, e^{-\frac{x^2}{2\sigma_X^2}},$$

a qualitative analysis, illustrated in Figure A.17, indicates that the output of the rectifier is folded so that the output signal has a distribution that occupies half the domain, but has double the amplitude so as to maintain the unitary area underneath the curve.

$$p_Y(y) = \frac{2}{\sigma_X \sqrt{2\pi}}\, e^{-\frac{y^2}{2\sigma_X^2}}\, u(y).$$

In addition to this, the Root Mean Square (RMS) value of the signal is reduced, because it is calculated as a function of the width of the pdf, which is reduced during the rectification. ■

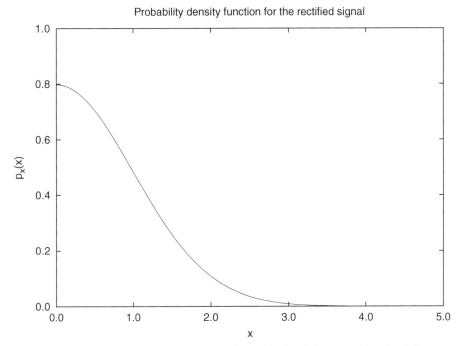

Figure A.17 Result for the distribution of a Gaussian signal that goes through a full-wave rectification.

Example: a method for building a random process is presented in the following. Consider the random variable ϕ with uniform distribution in the interval $[0, 2\pi]$, shown in Figure A.18.

The function $Y = V\cos\phi$ does not have an inverse for every point, therefore, it must be divided into parts to obtain the inverses.

$$p_Y(y) = \frac{p_\phi(\phi)}{|dy/d\phi|}, \qquad y = V\cos\phi \Rightarrow \phi = \cos^{-1}(y/V),$$

$$\frac{dy}{d\phi} = -V\sin\phi = -V\sqrt{1 - (y/V)^2} = -\sqrt{V^2 - y^2},$$

$$p_Y(y) = 2 \cdot \left[\frac{1}{2\pi} \cdot \frac{1}{\sqrt{V^2 - y^2}}\right] = \left[\frac{1}{\pi\sqrt{V^2 - y^2}}\right].$$

The result is illustrated in Figure A.19. It can be noticed that the distribution obtained represents the probability of the sine wave, for which the more

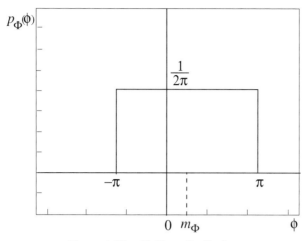

Figure A.18   Uniform distribution.

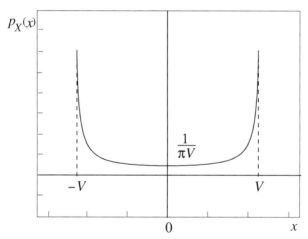

Figure A.19   The probability distribution function that is obtained from the mapping $y = V \cos \phi$.

extreme values have a greater probability of occurrence. A random process can be built merely by including the variable time in the discussed equation $Y = V \cos(\omega t + \phi)$. ∎

Furthermore, with regard to the pdf transformation, also called the function of a random variable, there are two situations of particular interest involving the uniform distribution:

First situation: transformation of a given pdf into a uniform pdf.

In this case, the output distribution is given by

$$p_y(y) = \frac{p_X(x)}{\left|\frac{dy}{dx}\right|}, \quad x = f_{-1}(y), \quad \frac{dy}{dx} \geq 0,$$

$$p_y(y) = \frac{p_X(x)}{dy/dx} \quad \Rightarrow \quad 1 = \frac{p_X(x)}{dy/dx}, \quad 0 \leq y \leq 1.$$

Therefore,

$$p_X(x) = \frac{dy}{dx} \quad \Rightarrow \quad dy = p_X(x)\,dx.$$

Then, integrating the differential in y, the desired function is obtained

$$f(x) = \int_{-\infty}^{y} dy = \int_{-\infty}^{x} p_X(x)\,dx = P_X(x).$$

The operation is useful, for example, for the analysis of signal compression and expansion systems, or for the study of non-uniform quantizers, which are used in telephony systems.

Second case: transformation of a uniform distribution into another given pdf.

Consider the calculation of the pdf of the output of the system,

$$p_Y(y) = \frac{p_X(x)}{dy/dx}, \quad p_X(x) = 1, \quad 0 \leq x \leq 1.$$

Therefore,

$$p_y(y) = \frac{1}{dy/dx} \quad \Rightarrow \quad \frac{dy}{dx} = \frac{1}{p_y(y)}, \quad y = g(x).$$

Solving the equation to obtain y, it is seen that the function required to transform the uniform distribution into another given distribution is the inverse of the CDF of the input, as follows

$$y = \int_{-\infty}^{x} \frac{1}{p_Y(g(x))}\,dy \quad \Rightarrow \quad y = P_X^{-1}(x).$$

This operation is useful to generate the random variables of any distribution in a computer, using a variable with uniform distribution, which is typically found in any computational language.

A.10 Discrete Distributions

The discrete probability distributions can be written, with the set of real numbers $\mathbb{R}$, using the generalized impulse function, or the Dirac Delta function.

The pdf for a random variable with Bernoulli distribution is given by

$$p_X(x) = p\delta(x + a) + q\delta(x - b), \tag{A.35}$$

illustrated in Figure A.20, and its CDF is

$$P_X(x) = pu(x + a) + qu(x - b), \tag{A.36}$$

which is illustrated in Figure A.21, in which $p + q = 1$.

In the case where levels a and b are equal, and the associated probabilities, p and q, are equal, then

$$p_X(x) = \frac{1}{2}\left[\delta(x + a) + \delta(x - a)\right]. \tag{A.37}$$

and its CDF is

$$P_X(x) = \frac{1}{2}\left[u(x + a) + u(x - a)\right]. \tag{A.38}$$

This distribution usually models data transmission in a communications system.

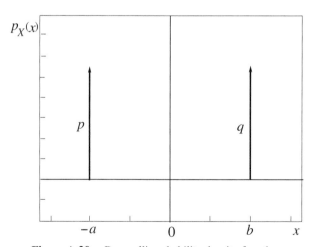

Figure A.20   Bernoulli probability density function.

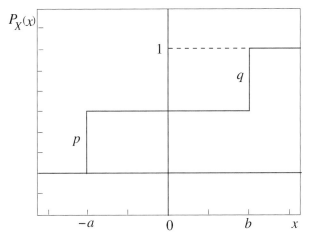

Figure A.21   Bernoulli cumulative distribution function.

The binomial pdf is a discrete distribution of the number of successes in a sequence of n independent Bernoulli trials. In each trial, the probability of success, p, is the same.

$$p_X(x) = \sum_{k=0}^{\infty} \binom{n}{k} p^k q^{n-k} \delta(x - k),\qquad (A.39)$$

and the CDF is written as

$$P_X(x) = \sum_{k=0}^{\infty} \binom{n}{k} p^k q^{n-k} u(x - k).\qquad (A.40)$$

The expected value of the binomial random variable is $E[X] = np$ and its variance is given by $V[X] = np(1 - p)$.

The geometric distribution, which has its pdf as follows, models the arrival of packets in a computer network, when the arrival and service rates are constant.

$$p_X(x) = \sum_{k=0}^{\infty} (1 - \rho)\rho^k \delta(x - k).\qquad (A.41)$$

The geometric pdf is shown in Figure A.22. The heights of the impulses are merely illustrative, and only indicate the relative areas.

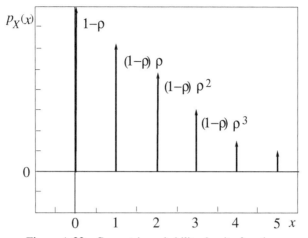

Figure A.22   Geometric probability density function.

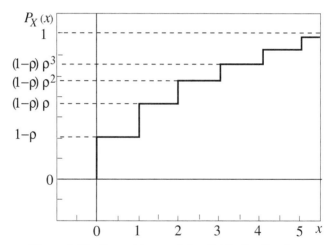

Figure A.23   Geometric cumulative distribution function.

The geometric CDF is given by

$$P_X(x) = \sum_{k=0}^{\infty}(1 - \rho)\rho^k u(x - k).\qquad\text{(A.42)}$$

The geometric CDF is illustrated in Figure A.23.

Example: calculate the mean value of a random variable with a geometric distribution.

According to Equation A.23, the mean value of a discrete random variable can be calculated by

$$E[X] = \sum_{k=0}^{\infty} p_k x_k$$

$$= \sum_{k=0}^{\infty} k(1 - \rho)\rho^k$$

$$= (1 - \rho) \sum_{k=0}^{\infty} k\rho^k.$$

Because the geometric series converges for $\rho < 1$,

$$\sum_{k=0}^{\infty} \rho^k = \frac{1}{1 - \rho},$$

and

$$\frac{d}{d\rho}\left(\frac{1}{1 - \rho}\right) = \frac{1}{(1 - \rho)^2},$$

and the derivative of a sum is the sum of the derivatives,

$$\frac{d}{d\rho}\left(\sum_{k=0}^{\infty} \rho^k\right) = \sum_{k=0}^{\infty} k\rho^{k-1} = \frac{1}{\rho} \sum_{k=0}^{\infty} k\rho^k,$$

then the expected value is

$$E[X] = \frac{\rho}{1 - \rho}. \blacksquare$$

The Poisson distribution, illustrated in Figure A.24, was discovered by Siméon Denis Poisson (1781–1840) and published in his work *Recherches sur la Probabilité des Jugements en Matières Criminelles et Matière Civile*, in which his probability theory also appears, in 1838.

The Poisson pdf is given by

$$p_X(x) = \sum_{k=0}^{\infty} \frac{e^{-\lambda}\lambda^k}{k!}\delta(x - k), \tag{A.43}$$

and the CDF is

$$P_X(x) = \sum_{k=0}^{\infty} \frac{e^{-\lambda}\lambda^k}{k!} u(x - k). \tag{A.44}$$

The Poisson CDF is illustrated in Figure A.25. This distribution is commonly used to describe situations like particle emissions from radioactive materials, telephone traffic, and even the formation of lines in banks. The

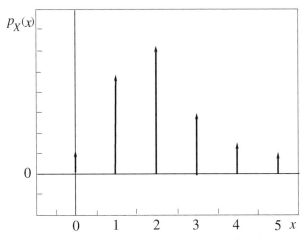

Figure A.24   Poisson probability density function.

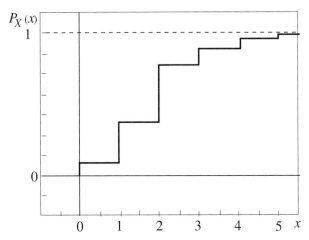

Figure A.25   Poisson cumulative distribution function.

expected value of a Poisson variable and its variance are both equal to the distribution parameter λ.

An important aspect to consider with discrete probability distributions is that, for a finite, or countable, number of points, it is possible to assign nonzero probability values for all the points in the set. One may notice that, for a non-countable set of points, that is, for continuous distributions, it is impossible to assign nonzero probabilities to all points of the set. That is why the pdf is used when dealing with continuous variables, instead of sequences of probabilities, which results in the need to use the impulse function to model the discrete case.

A.11 Characteristic Function

The characteristic function is the Fourier transform of the pdf. It was created by Dirac as a way to solve the problem of the distribution of a sum of random variables.

$$P_X(\omega) = \mathrm{E}[\mathrm{e}^{-j\omega X}] = \int_{-\infty}^{\infty} \mathrm{e}^{-j\omega x} \, p_X(x) \, \mathrm{d}x. \qquad (A.45)$$

For discrete probability distributions, like Equation A.22, the characteristic function can be calculated as

$$
\begin{aligned}
P_X(\omega) &= \int_{-\infty}^{\infty} \mathrm{e}^{-j\omega x} \sum_{k=-\infty}^{\infty} p_k \delta(x - x_k) \, \mathrm{d}x. \\
&= \sum_{k=-\infty}^{\infty} p_k \int_{-\infty}^{\infty} \mathrm{e}^{-j\omega x} \, \delta(x - x_k) \, \mathrm{d}x. \\
&= \sum_{k=-\infty}^{\infty} p_k \, \mathrm{e}^{-j\omega x_k}, \qquad (A.46)
\end{aligned}
$$

which is the discrete Fourier transform of the distribution, or sequence of probabilities $\{p_k\}$.

The characteristic function is obtained by making $g(X) = \mathrm{e}^{-j\omega X}$ in the general expression

$$\mathrm{E}[g(X)] = \int_{-\infty}^{\infty} g(x) p_X(x) \, \mathrm{d}x.$$

If the random variable X is given in volts, then $p_X(x)$ is given in $[V^{-1}]$ and $P_X(\omega)$, the characteristic function, is always dimensionless. The unit for ω is the inverse of the unit for the random variable.

The characteristic function can also be used to find the moments of a random variable, provided the n-th moment exists. Consider the first and second-order moments of the random variable X, given by

$$E[X] = \int_{-\infty}^{\infty} x p_X(x)\, dx$$

and

$$E[X^2] = \int_{-\infty}^{\infty} x^2 p_X(x)\, dx.$$

Differentiating Equation A.45,

$$\frac{\partial}{\partial\omega} P_X(\omega) = \frac{\partial}{\partial\omega} \int_{-\infty}^{\infty} e^{-j\omega x}\, p_X(x)\, dx = \int_{-\infty}^{\infty} (-j)x e^{j\omega x}\, p_X(x)\, dx.$$

Making $\omega = 0$ in the equation yields

$$\frac{1}{-j} \cdot \frac{\partial}{\partial\omega} P_X(\omega)\bigg|_{\omega=0} = \int_{-\infty}^{\infty} x\, p_X(x)\, dx = E[X],$$

which is the expected value of X.

The second derivative produces the second moment,

$$\frac{1}{(-j)^2} \cdot \frac{\partial^2}{\partial\omega^2} P_X(\omega)\bigg|_{\omega=0} = \int_{-\infty}^{\infty} x^2 p_X(x)\, dx = E[X^2].$$

Therefore, all the moments of the random variable X can be obtained from the characteristic functions, through successive differentiation

$$E[X^n] = \frac{1}{(-j)^n} \cdot \frac{\partial^n}{\partial\omega^n} P_X(\omega)\bigg|_{\omega=0}. \tag{A.47}$$

Example: consider the distribution $p_X(x) = \alpha\, e^{-\alpha x}\, u(x)$. Its characteristic function is calculated as follows

$$P_X(\omega) = \int_{-\infty}^{\infty} e^{-j\omega x}\, p_X(x)\, dx = \int_{0}^{\infty} e^{-j\omega x} \cdot \alpha \cdot e^{-\alpha x}\, dx$$

$$= \alpha \int_{0}^{\infty} e^{-x(j\omega + \alpha)}\, dx,$$

$$P_X(\omega) = \frac{\alpha}{(\alpha + j\omega)} e^{x(j\omega + \alpha)} \Big|_0^\infty = \frac{\alpha}{(\alpha + j\omega)}.$$

The first moment, or the mean of the random variable, is determined by calculating the first derivative of the characteristic function

$$\mathrm{E}[X] = \frac{1}{-j} \cdot \frac{\partial P_X(\omega)}{\partial \omega}\Big|_{\omega=0} = \frac{1}{-j}\left[\frac{-\alpha \cdot (j)}{(\alpha + j\omega)^2}\right]\Big|_{\omega=0} = \frac{1}{\alpha}.$$

The result can be verified by computing the mean by the usual formula. ∎

A.12 Conditional Distribution

A probability distribution can be conditioned, by using the Bayes formula for the conditional probability between sets, and defining the sets accordingly

$$P(A|B) = \frac{P(A \cap B)}{P(B)}.$$

Set A is defined as $A = \{X \leq x\}$, to make it possible to calculate the CDF

$$P\{X \leq x\} = P_X(x).$$

Set B can be defined using any line segment obtained from the Borel family, which is the family of all subsets from the real set $\mathbb{R}$ that can be represented by semi-open line segments, that is, $B = \{a < X \leq b\}$, as shown in Figure A.26.

The probability measure of this set is given by

$$P\{a < X \leq b\} = P_X(b) - P_X(a).$$

Example: suppose that $B = (-\infty, a]$, as indicated in Figure A.27. The conditional probability is then given by

$$P\{X \leq x|B\} = \frac{P\{X \leq x, -\infty < X \leq a\}}{P\{-\infty < X \leq a\}}.$$

Considering that the intersection of the segments for $X \leq a$ is the set $\{X \leq x\}$ and that the intersection for $X > a$ is $\{X \leq a\}$, then

$$P_X(x|B) = P\{X \leq x|B\} = \begin{cases} \frac{P_X(x)}{P_X(a)}, & x \leq a \\ \frac{P_X(a)}{P_X(a)} = 1, & x > a \end{cases}.$$

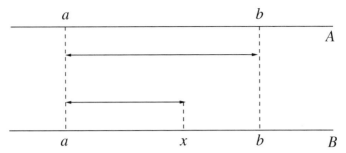

Figure A.26   Line segments used in conditional probability.

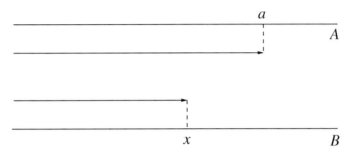

Figure A.27   Line segments used in the example.

The conditional pdf is obtained by differentiating the conditional CDF

$$p_X(x|B) = \frac{d}{dx} P_X(x|B) = \begin{cases} \frac{p_X(x)}{P_X(a)}, & X \leq a \\ 0, & X > a \end{cases}.$$

The two functions are in Figure A.28 and Figure A.29, in which the dashed line indicates the conditional distributions. One may notice that conditioning a CDF is equivalent to providing information about the random variable.

The knowledge that the variable is, *a priori*, in a given interval elevates the CDF, as well as the pdf. This can easily be seen, because the denominator is always less than one unless the interval is the set of real numbers itself. ■

Example: consider an interval $B = [a, b]$. The conditional distribution is obtained as was done in the previous calculations and is illustrated in Figure A.30,

Example: consider an interval $B = [a, b]$. The conditional distribution is obtained as was done in the previous calculations, and is illustrated in

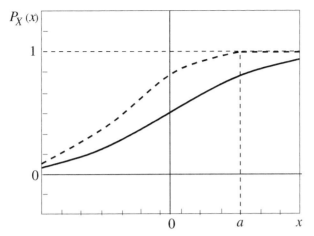

Figure A.28   Conditional cumulative distribution function.

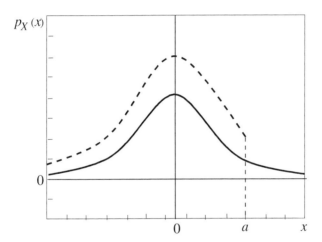

Figure A.29   Conditional probability density function.

Figure A.30,

$$P_X(x|B) = \frac{P\{X \le x, a < X \le b\}}{P\{a < X \le b\}},$$

in which, by definition,

$$P\{a < X \le b\} = P_X(b) - P_X(a).$$

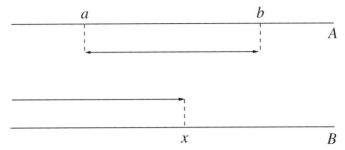

Figure A.30   Line segments used to calculate the conditional probability.

The conditional CDF is given by

$$P_X(x|B) = \begin{cases} 0 & , \quad X \le a \\ \frac{P_X(x) - P_X(a)}{P_X(b) - P_X(a)} & , \quad a < X \le b \\ 1 & , \quad X > b \end{cases}$$

and the conditional pdf can be immediately obtained as

$$P_X(x|B) = \begin{cases} 0 & , \quad X \le a \\ \frac{p_X(x)}{P_X(b) - P_X(a)} & , \quad a < X \le b \\ 0 & , \quad X > b. \end{cases}$$

The conditional cumulative probability distribution is shown in Figure A.31, and the conditional probability density function is depicted in Figure A.32. In both graphs, the contiguous line represents the original distributions, and the dashed line indicates the conditional distributions. ∎

Example: consider Figure A.33, which represents a channel with additive noise, commonly used to model communications systems.

The transmitted signal initially assumes value A. The additive signal, X, that can be the noise, is usually independent of the transmitted signal.

This problem can be solved using conditional probability, or by the pdf transformation. Using the second approach leads to

$$p_Y(y) = \frac{p_X(x)}{|dy/dx|}, \quad |dy/dx| = 1, \quad x = y - A$$

$$p_Y(y) = p_X(y - A)$$

Using conditional probability yields

$$P_Y(y|A) = \frac{P_Y(y, A)}{P(A)} = \frac{P_X(y - A) \cdot P(A)}{P(A)}$$

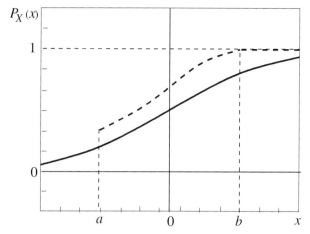

Figure A.31   Conditional cumulative probability distribution.

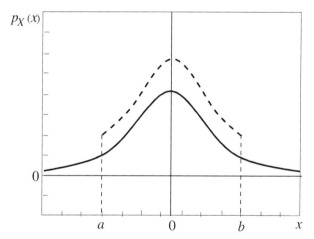

Figure A.32   Conditional probability density function.

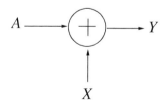

Figure A.33   Additive noise channel, $Y = A + X$.

Simplifying the expression and calculating its derivative yields the result previously obtained.

For the case where the input is $-A$ or A, then

$$p_Y(y) = \frac{1}{2} p_X(y+A) + \frac{1}{2} p_X(y-A).$$

Note that the noise causes an intersection between the pdf curves. This interference is responsible for the detection errors in the received signal. The error probability is obtained using Bayes Rule,

$$P_e = P(e|A)P(A) + P(e|-A)P(-A),$$

in which $P(e|A)$ is the error probability given that the pulse with value A was transmitted, and $P(e|-A)$ represents the error probability in the case where $-A$ was transmitted.

Increasing the signal power improves detection, as it causes a separation between the conditional pdf curves. The transmitted signal power is A^2. ■

Example: for the specific case of an Additive White Gaussian Noise (AWGN) channel, the formulas become

$$p_N(n) = \frac{1}{\sigma_N \sqrt{2\pi}} \, e^{-\frac{N^2}{2\sigma_N^2}}$$

and

$$p_Y(y) = \frac{1}{2} \frac{1}{\sigma_N \sqrt{2\pi}} \, e^{-\frac{(n+A)^2}{2\sigma_N^2}} + \frac{1}{2} \cdot \frac{1}{\sigma_N \sqrt{2\pi}} \, e^{-\frac{(n-A)^2}{2\sigma_N^2}} .■$$

Gaussian noise is common in communications systems, in which it typically occurs as thermal noise, in control systems, in which it appears as the estimation error of the controller, and in power systems, in which it models the effects of harmonics caused by thyristor-based equipment. The Gaussian error is also important to characterize population statistics.

A.13 Useful Distributions and Applications

A well-accepted model used for signal amplitude statistics in an environment subject to fading uses the Rayleigh distribution, named after John William Strutt (1842–1919), English mathematician and physicist, also known as Baron of Rayleigh who has done extensive research on undulatory phenomena, also known as Baron of Rayleigh (Kennedy, 1969).

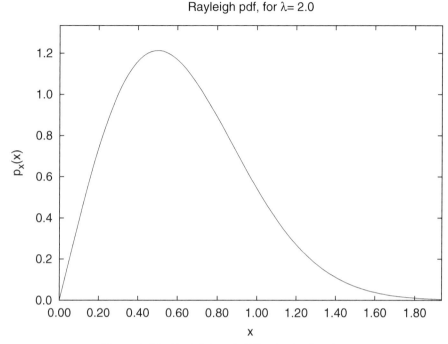

Figure A.34   Rayleigh probability density function.

This distribution, shown in Figure A.34, represents the effect of multiple signals, reflected and refracted when detected by the receiver, with no direct line of sight, or direct ray (Lecours et al., 1988).

The Rayleigh pdf illustrated in Figure A.34 for different parameter values, is given by (Proakis, 1990)

$$p_X(x) = \frac{x}{\sigma^2} e^{-\frac{x^2}{2\sigma^2}}, u(x) \tag{A.48}$$

with mean $E[X] = \sigma\sqrt{\pi/2}$ and variance $V[X] = (2 - \pi)\frac{\sigma^2}{2}$. It's CDF is shown in Figure A.35.

In this case, the phase distribution can be considered uniform for the interval $(0, 2\pi)$. It can be observed that a distribution equivalent to Rayleigh can be obtained with only six waves, with independently distributed phases (Schwartz et al., 1966).

If a strong main component is considered, that is, if the previously described scenario has a line of sight between transmitter and receiver, in

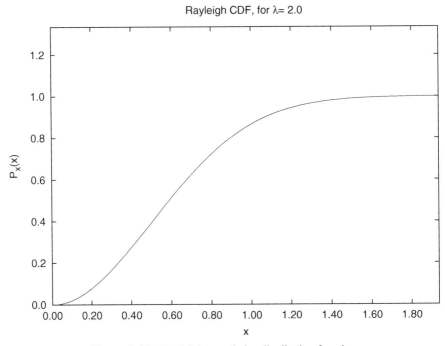

Figure A.35 Rayleigh cumulative distribution function.

addition to the multipath components, the Rice distribution, named after a pioneer in the fields of Communication and Information Theory, Stephen Oswald Rice (1907–1986), can be used to describe the fast variations in the signal envelope. This line of sight component reduces the variance in the signal amplitude, as it increases in comparison to the multipath components (Lecours et al., 1988) (Rappaport, 1989). The Rician distribution is given by

$$p_X(x) = \frac{x}{\sigma^2} e^{-\frac{x^2+A^2}{2\sigma^2}} I_0\left(\frac{xA}{\sigma^2}\right) u(x), \tag{A.49}$$

in which $I_0(\cdot)$ is the zero-order Bessel function, written as

$$I_0(x) = \frac{1}{\pi} \int_0^{\pi} e^{x \cos \theta} d\theta. \tag{A.50}$$

and A is the signal amplitude. The variance, for a unitary mean, is $V[X] = A^2 + 2\sigma^2 + 1$. And the expected value is given by

$$E[X] = e^{-\frac{A^2}{4\sigma^2}} \sqrt{\frac{\pi}{2}} \sigma [(1 + \frac{A^2}{2\sigma^2}) I_0(\frac{A^2}{4\sigma^2}) + \frac{A^2}{2\sigma^2} I_1(\frac{A^2}{4\sigma^2})], \qquad (A.51)$$

in which $I_1(\cdot)$ is the modified first-order Bessel function, written as

$$I_1(x) = \frac{1}{\pi} \int_0^\pi e^{x \cos \theta} \cos(\theta) d\theta. \qquad (A.52)$$

The term $A^2/2\sigma^2$ is a measure of the fading statistics. As $A^2/2\sigma^2$ increases, the effect of the multiplicative noise, or fading, becomes less important. The pdf of the signal becomes more concentrated around the main component. The remaining disturbances are perceived as phase fluctuations.

On the other hand, the signal weakening can cause the main component to remain undetected among the multipath components, which results in the Rayleigh model. And increasing A in relation to the standard deviation, σ, makes the statistics converge to a Gaussian distribution with mean A (Schwartz, 1970).

Another distribution that is useful when modeling multipath fading scenarios is the Nakagami pdf. This distribution can be applied in cases where there is a random superposition of random vectorial components (Neal H. Shepherd, Editor, 1988). The Nakagami distribution can be expressed as

$$p_X(x) = \frac{2m^m x^{2m-1}}{\Gamma(m)\Omega^m} e^{-\frac{mx^2}{\Omega}} u(x), \qquad (A.53)$$

in which $\Omega = P_X$ is the mean power of the received signal, $m = \Omega^2/E[(X - E[X])^2]$ represents the inverse of the normalized variance of X^2 and $\Gamma(\cdot)$ is the Gamma function. The parameter m is known as the distribution modeling factor and cannot be less than $1/2$.

It can be shown that the Nakagami pdf represents a more general expression, that covers many other distributions. For example, for $m = 1$ the Rayleigh distribution is obtained.

The lognormal distribution is used to model the effect of certain topography patterns on the transmitted signal, that appear because of the non-homogeneity of the channel or due to transmission in very obstructed or congested environments (Hashemi, 1991; Rappaport, 1989).

The lognormal distribution, which can also model communication channels in industrial environments, is represented by

$$p_R(r) = \frac{1}{\sigma r \sqrt{2\pi}} e^{-\frac{(\log r - \mu)^2}{2\sigma^2}} u(r), \qquad (A.54)$$

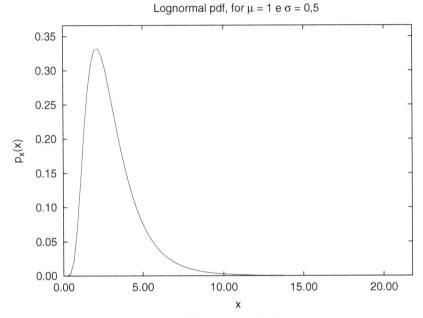

Figure A.36   Lognormal pdf.

which has mean value $E[X] = e^{\sigma^4/2+\mu}$, and variance $V[X] = e^{\sigma^4+\mu}$ $(e^{\sigma^4} - 1)$.

It can be obtained directly from the Gaussian distribution by the use of an appropriate transformation. Its pdf is shown in Figure A.36, for $\mu = 1$ and $\sigma = 0,5$.

The CDF of the lognormal distribution is illustrated in Figure A.36. Note that the function increases faster than the Gaussian CDF.

The von Mises distribution was introduced in 1918, by Richard Edler von Mises (1883–1953), an Austrian mechanical engineer and mathematician, for the study of deviations in atomic weights in relation to integer values (Mises, 1918). Recently, it had an important role in the modeling and statistical analysis of angular variables. Consider the random variable Θ representing the angle of arrival of a certain signal, from a multipath component (scattered and specular). The von Mises distribution for the scattered component of Θ is given by

$$p_\Theta(\theta) = \frac{1}{2\pi I_0(\kappa)} e^{\kappa \cos(\theta - \theta_M)}, \quad \theta \in [-\pi, \pi), \qquad (A.55)$$

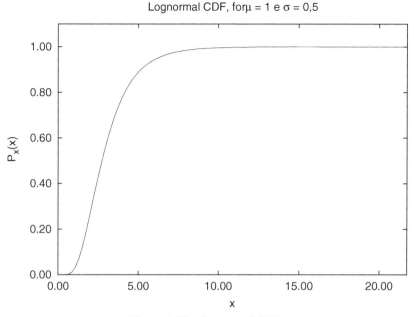

Figure A.37 Lognormal CDF.

in which $I_0(\cdot)$ is the modified zero-order Bessel function and $\theta_M \in [-\pi, \pi)$ represents the mean direction for the angle of arrival of the scattered components and $\kappa \geq 0$ controls the width of the arrival angle.

The distribution discovered by the Italian civil engineer and economist Vilfredo Federico Damaso Pareto (1848–1923) is defined as (Pareto, 1964)

$$p(x) = \frac{\alpha \beta^\alpha}{x^{\alpha+1}}, \quad \alpha, \beta > 0. \tag{A.56}$$

The CDF is

$$P(x) = 1 - \left(\frac{\beta}{x}\right)^\alpha, \quad x > \beta. \tag{A.57}$$

The first moment, or expected value is given by

$$m_1 = E[x] = \frac{\alpha \beta}{(\alpha - 1)}, \quad \alpha > 1. \tag{A.58}$$

The distribution was discovered by Waloddi Weibull (1887–1979), a Swedish engineer and mathematician is defined as (Weibull, 1951)

$$p(x) = \alpha x^{\alpha-1} \beta^{-\alpha} e^{-\left(\frac{x}{\beta}\right)^\alpha}, \quad \alpha, \beta > 0. \tag{A.59}$$

The CDF is

$$P(x) = 1 - e^{-\left(\frac{x}{\beta}\right)^{\alpha}}.$$ (A.60)

And the first moment, or expected value is written as

$$m_1 = E[x] = \beta\Gamma\left(1 + \frac{1}{\alpha}\right),$$ (A.61)

and the variance is calculated as

$$\text{Var}[x] = \sigma^2 = \beta^2\left(\Gamma\left(1 + \frac{2}{\alpha}\right) - \Gamma\left(1 + \frac{1}{\alpha}\right)^2\right).$$ (A.62)

A.14 Joint Random Variables

A.14.1 An Extension of the Concept of Random Variables

A joint random variable represents a mapping of joint random events, defined as a result of operations carried out in an algebra of sets. It is a transposition of events of the sample space Ω, to Borel rectangles, in the set of real numbers.

The way the mathematician Félix Borel structured the family of sets, or events, that bear his name, permits a simple generalization of the probability measure for any dimension.

The joint random variables form a natural extension of the concept of random variables. For two variables, the regions become rectangles in a wider sense. The following examples illustrate two-dimensional Borel sets, that is, for $\Omega = \mathbb{R}^2$.

Example: the rectangle $R = \{-\infty < X \leq x, -\infty < Y \leq y\}$ is illustrated in Figure A.38. Calling this region R, the two-dimensional cumulative function can be defined as

$$P_{XY}(x, y) = P(R) = P\{-\infty < X \leq x, -\infty < Y \leq y\}.\blacksquare$$ (A.63)

The definition of a random variable can be extended to a multidimensional space. The real functions $X_1, X_2, \ldots, X_n$ are random variables if and only if $\forall x_1, x_2, \ldots, x_n \in \mathbb{R}$, the following relations are satisfied $\{\omega : X_1(\omega) \leq x_1, X_2(\omega) \leq x_2, \ldots, X_n(\omega) \leq x_n\} \in \mathcal{F}$.

If f is an n-dimensional Borelian function defined in the considered space, and $X_1, X_2, \ldots, X_n$ are random variables, also called a vector, then the function $f(X_1, X_2, \ldots, X_n)$ is also a random variable.

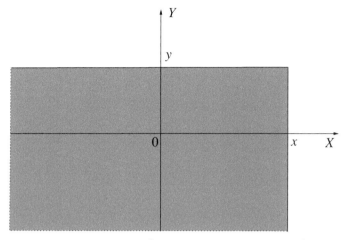

Figure A.38   Region $\{-\infty < X \le x, \ -\infty < Y \le y\}$.

To prove this statement, it is sufficient to consider the previous vector $X = X_1, X_2, \ldots, X_n$. Using the property of inverse functions, it can be shown that $f^{-1}(X)(\mathcal{B}(\mathbb{R})) = X^{-1}\left[f^{-1}(X)(\mathcal{B}(\mathbb{R}))\right] \subset X^{-1}(\mathcal{B}(\mathbb{R}^n)) \in \mathcal{F}$.

The two-dimensional Cumulative Distribution Function (CDF) is the base for the definition of other mathematical constructions, as long as they are constructed from a measurable space $(\Omega, \mathcal{F}, \mathcal{P})$, in which the universe set is the real plane, $\Omega = \mathbb{R}^2$, $\mathcal{F} = \mathcal{B}\,(\mathbb{R}^2)$ represents the family of Borel rectangles and P is the measure of probability.

Example: the region $R_1 = \{-\infty < X \le x_1, \ y_1 < Y \le y_2\}$, shown in Figure A.39, also defines a rectangle. It can be shown that the measure of that region can be expressed as a function of the joint CDF,

$$P\{-\infty < X \le x_1, \ y_1 < Y \le y_2\} = P_{XY}(x, y_2) - P_{XY}(x, y_1). \blacksquare \quad (A.64)$$

The following example shows how to obtain the pdf from the CDF in two dimensions, as long as infinitesimal rectangles, whose measures converge in the limit, are defined.

Example: the region $R_2 = \{x_1 < X \le x_2, \ y_1 \le Y \le y_2\} = R_2$ has probability measure given by

$$P(R_2) = P_{XY}(x_2, y_2) - P_{XY}(x_1, y_2) - P_{XY}(x_2, y_1) + P_{XY}(x_1, y_1). \blacksquare$$
$$(A.65)$$

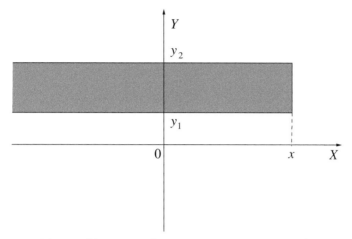

Figure A.39   Region $\{-\infty < X \le x_1,\, y_1 < Y \le y_2\}$.

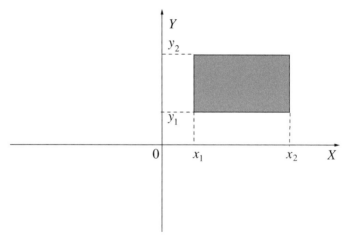

Figure A.40   Region $\{x_1 < X \le x_2,\, y_1 \le Y \le y_2\} = R_2$.

Using the preceding expression, and following the same steps taken in the study of unidimensional variables, the relations between the joint pdf and CDF can be obtained.

Expression A.65 is useful to compute the two-dimensional pdf. Defining $\Delta x = x_2 - x_1$ and $\Delta y = y_2 - y_1$, the pdf of the two-dimensional variable indicated in Figure A.40 can be obtained, dividing the probability $P(R_2)$ by the area of the rectangle, that is,

$$p_{XY}(x,y) = \frac{P_{XY}(x_1 + \Delta x, y_1 + \Delta y) - P_{XY}(x_1, y_2 + \Delta y)\Delta x \Delta y}{}$$

$$- \frac{P_{XY}(x_1 + \Delta x, y_1) - P_{XY}(x_1, y_1)}{\Delta x \Delta y}. \tag{A.66}$$

Taking the limits $\Delta x \to 0$ and $\Delta y \to 0$, the two-dimensional pdf is obtained as

$$p_{XY}(x,y) = \frac{\partial^2 P_{XY}(x,y)}{\partial x \partial y}. \tag{A.67}$$

The CDF is naturally obtained with the Fundamental Theorem of Calculus, which results in

$$P_{XY}(x,y) = \int_{-\infty}^{x} \int_{-\infty}^{y} p_{XY}(\alpha, \beta)\, d\alpha\, d\beta. \tag{A.68}$$

Example: the Rician distribution is used to model communication channels that present multiple paths, with a line of sight between the transmitter and the receptor.

The joint distribution, for amplitude and phase, which is the origin of the Rician model for amplitude variations, is given by

$$p_{X\Theta}(x,\theta) = \frac{x e^{-A^2/2}}{2\pi\sigma^2} e^{-(x^2 - 2xA\cos\theta)/2\sigma^2}, \ 0 \le x < \infty, \ -\pi \le \theta \le \pi. \tag{A.69}$$

Integrating the second-order cumulative function, in relation to the phase, θ, the marginal distribution, or first-order Rician pdf, is obtained,

$$p_X(x) = \frac{x}{\sigma^2} e^{-\frac{x^2 + A^2}{2\sigma^2}} I_0\left(\frac{xA}{\sigma^2}\right) u(x),$$

in which $I_0(\cdot)$ is the modified first-order Bessel function, given by Formula A.50.

Integrating the formula for all values of x yields the marginal distribution for the phase θ,

$$p_\Theta(\theta) = \frac{e^{-s^2}}{2\pi} + \frac{1}{2}\sqrt{\frac{s^2}{\pi}} \cos\theta\, e^{-s^2 \sin^2\theta}[1 + 2(1 - Q(s/\sqrt{2}))\cos\theta]. \blacksquare \tag{A.70}$$

This expression generates a bell-shaped curve for large values of the signal-to-noise ratio (SNR). For $A = 0$ the distribution converges to uniform

(Schwartz, 1970). The SNR is an auxiliary parameter given by $s^2 = A^2/2\sigma^2$. The function $Q(x)$ is defined in the usual form

$$Q(x) = \frac{1}{\sqrt{2\pi}} \int_x^\infty e^{\frac{-y^2}{2}} \, dy. \tag{A.71}$$

A.15 Properties of Probability Distributions

The joint probability density function and the joint cumulative probability function inherit the following properties from the Lebesgue measure:

1. The cumulative function is non-negative: $P_{XY}(x, y) \geq 0$.
2. The probability density function is also non-negative: $p_{XY}(x, y) \geq 0$.
3. The probability of an empty set is zero: $P_{XY}(-\infty, -\infty) = P(\emptyset) = 0$.
4. The probability of the universal set is one: $P_{XY}(-\infty, \infty) = P(\Omega) = 1$.
5. The marginal density in y is calculated integrating the joint distribution for x, $p_Y(y) = \int_{-\infty}^{\infty} p_{XY}(x, y) \, dx$.
6. The marginal density in x is calculated integrating the joint distribution for y, $p_X(x) = \int_{-\infty}^{\infty} p_{XY}(x, y) \, dy$.

Example: the Gaussian two-dimensional pdf is given by the formula

$$p_{XY}(x, y) = \frac{1}{2\pi\sigma_X\sigma_Y\sqrt{1-\rho^2}} e^{-\frac{1}{1-\rho^2}\left[\frac{(x-m_X)^2}{\sigma_X^2} - \frac{2\rho(x-m_X)(y-m_Y)}{\sigma_X\sigma_Y} + \frac{(y-m_Y)^2}{\sigma_y^2}\right]}, \tag{A.72}$$

in which ρ represents the correlation coefficient, m_X and m_Y are the respective means for the random variables X and Y, σ_X and σ_Y are the respective standard deviations. ∎

A.16 Moments in Two Dimensions

For the random variables X and Y defined in the probability space $(\mathbb{R}, \mathcal{B}(\mathbb{R}), P)$, their joint moments of order $n + m$ can be defined with the use of the Riemann integral,

$$E[X^n Y^m] = \int_{-\infty}^{\infty} \int_{-\infty}^{\infty} x^n y^m p_{XY}(x, y) \, dx \, dy, \quad \text{order} : n + m. \tag{A.73}$$

The statistical expectancy of a given function, $f(x, y)$ of the variables X and Y is defined by

$$E[f(x, y)] = \int_{-\infty}^{\infty} \int_{-\infty}^{\infty} f(x, y) p_{XY}(x, y)\, dx\, dy. \quad (A.74)$$

For the random variables X and Y, defined in the probability space of real numbers, the covariance, which is a measure of the correlation between variables, is defined as

$$C[X, Y] = E[(X - m_X)(Y - m_Y)] = E[XY] - E[X] \cdot E[Y]. \quad (A.75)$$

When there is no correlation between the variables, then $C[X, Y] = 0$

The correlation coefficient, ρ_{XY}, or simply ρ, is the covariance normalized by the product of the standard deviations

$$\rho_{XY} = \frac{C[X, Y]}{\sigma_X \sigma_Y}. \quad (A.76)$$

The correlation coefficient is limited to the interval $[-1, 1]$, that is, $-1 \leq \rho_{XY} \leq 1$, and the variables are said to be uncorrelated when $\rho_{XY} = 0$. When $\rho_{XY} \approx 1$, it is said that the variables are positively and strongly correlated, In the case in which $\rho_{XY} \approx -1$, the variables are negatively correlated.

Figures A.41, A.42, and A.43 illustrate the regions defined by different correlation coefficients and expected values. The standard deviations of X and Y also distort the shape of the regions.

The correlation between X and Y is a measure of the orthogonality between the variables, which is sometimes confusing. The correlation is given by the expected mean of the product between the random variables

$$R[X, Y] = E[XY]. \quad (A.77)$$

The correlation can be written as

$$R[X, Y] = C[X, Y] + E[X] \cdot E[Y], \quad (A.78)$$

and, for two random variables to be uncorrelated, the correlation must be equal to the product of the means, that is,

$$R[X, Y] = E[X] \cdot E[Y] = m_X \cdot m_Y.$$

The correlation has some useful properties that help establish limits for calculations. For example, it can be related to the sum of the second moments of X and Y,

$$R[X, Y] \leq \frac{E[X^2] + E[Y^2]}{2}, \quad (A.79)$$

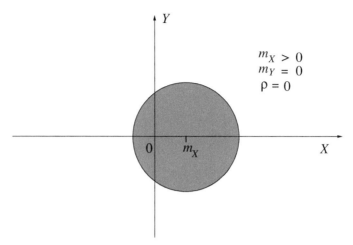

Figure A.41   Region defined by a zero correlation coefficient.

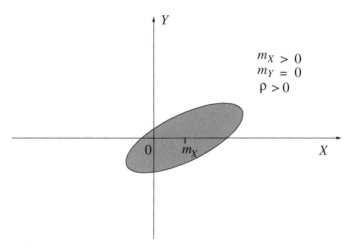

Figure A.42   Region defined by a positive correlation coefficient.

or can be related to the product of the second moments of X and Y,

$$\mathrm{R}[X,Y] \leq \sqrt{\mathrm{E}[X^2] \cdot \mathrm{E}[Y^2]}. \tag{A.80}$$

To demonstrate the second property it suffices to use the following tautology, which is always true,

$$\mathrm{E}[(\alpha X - Y)^2] \geq 0.$$

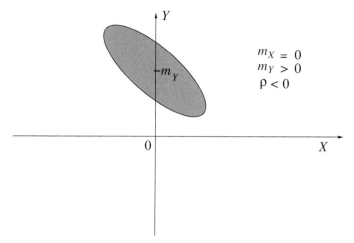

$$m_X = 0$$
$$m_Y > 0$$
$$\rho < 0$$

Figure A.43   Region defined by a negative correlation coefficient.

Calculating the square of the binomial yields

$$E[\alpha^2 X^2 - 2\alpha XY + Y^2] \geq 0.$$

Using the linearity of the expected value,

$$\alpha^2 E[X^2] - 2\alpha E[XY] + E[Y^2] \geq 0,$$

that results in

$$\alpha^2 E[X^2] - 2\alpha R[X, Y] + E[Y^2] \geq 0,$$

which is an inequality in terms of the variable α, while the expected values act as parameters for the inequality.

For the parabola to be above, or at least, touching the X-axis, the discriminant of the equation must be negative, that is,

$$\Delta = 4R^2[X, Y] - 4E[X^2] \cdot E[Y^2] \leq 0.$$

Therefore, after simplifying, one obtains

$$R^2[X, Y] \leq E[X^2] \cdot E[Y^2],$$

and the property follows. To demonstrate the first property, is suffices to attribute $\alpha = 1$ and repeat the deduction.

Example: it is important to note that the lack of correlation does not imply that the variables are independent, which is a more strict condition, and usually difficult to prove. Suppose that

$$X = V \cos \theta$$

and

$$Y = V \sin \theta,$$

in which θ is a random variable with uniform distribution in the interval $[0, 2\pi]$, and V is a constant.

The means of X and Y are zero, because

$$E[X] = E[V \cos \theta] = \int_{-\infty}^{\infty} V \cos \theta \frac{1}{2\pi} [u(\theta + \pi) - u(\theta - \pi)] d\theta$$

$$= \frac{V}{2\pi} \int_0^{2\pi} \cos \theta d\theta = \frac{V}{2\pi} [\sin \theta]_0^{2\pi} = 0. \tag{A.81}$$

and

$$E[Y] = E[V \sin \theta] = \frac{V}{2\pi} \int_0^{2\pi} \sin \theta d\theta = \frac{V}{2\pi} [\cos \theta]_0^{2\pi} = 0.$$

The correlation between the variables is zero, which is shown as follows

$$E[XY] = V^2 E[\cos \theta \cdot \sin \theta] = \frac{V^2}{2\pi} \int_0^{2\pi} \cos \theta \sin \theta d\theta = 0,$$

because the sine and cosine are orthogonal functions.

However, the sum of squares of X and Y results in $X^2 + Y^2 = V^2$, indicating that the variables are dependent. ■

The variance of the sum of the uncorrelated variables X and Y is given by

$$\begin{aligned}
V[X + Y] &= E\left[((X + Y) - E[X + Y])^2\right] \\
&= E\left[(X + Y)^2\right] - (E[X] + E[Y])^2 \\
&= E\left[X^2\right] - E^2[X] + E\left[Y^2\right] - E^2[Y] \\
&\quad + 2\left(E[XY] - E[X] \cdot E[Y]\right) \\
&= V[X] + V[Y], \tag{A.82}
\end{aligned}$$

because $E[XY] = E[X] \cdot E[Y]$, for uncorrelated variables.

Example: if the correlation coefficient is null, $\rho = 0$, the joint probability density function simplifies to

$$p_{XY}(x, y) = \frac{1}{2\pi\sigma_X\sigma_Y} e^{-\left[\frac{(x - m_X)^2}{\sigma_X^2} + \frac{(y - m_y)^2}{\sigma_y^2}\right]},$$

which can be written as the product of the marginal distributions,

$$p_{XY}(x,y) = \frac{1}{\sigma_X\sqrt{2\pi}} e^{-\frac{(x-m_X)^2}{\sigma_X^2}} \cdot \frac{1}{\sigma_Y\sqrt{2\pi}} e^{-\frac{(y-m_Y)^2}{\sigma_y^2}} = p_X(x) \cdot p_Y(y),$$

that is, decorrelation implies independence for Gaussian random variables. ∎

A.17 Conditional Moments

For the random variables X and Y, defined in the probability space $(\mathbb{R}, \mathcal{B}((\mathbb{R}), P)$, the conditional expectancy of X given $Y = y$, is defined using the Stieltjes integral, as

$$E[X|Y = y] = \int_{-\infty}^{\infty} x\, dP_{X|Y}(x|Y = y), \tag{A.83}$$

in which $P_{X|Y}(x|Y = y)$ is the conditional probability distribution of X given $Y = y$.

One must remember that $E[X|Y = y]$ is a function of y, that can be written as $E[X|y]$, known in Statistics as the regression of X in relation to Y. It represents the expectancy of X given the information about the occurrence of Y.

This concept is important for the definition of conditional entropy, mutual information, and channel capacity. In case the value of Y is not specified, then $E[X|Y]$ is a function of the random variable Y, for which the expected value is $E[E[X|Y]] = E[X]$, demonstrated as follows.

Because the variables are continuous, which allows the use of the Riemann integral, and the joint pdf is known, then

$$\begin{aligned} E[X|Y = y]] &= \int_{-\infty}^{\infty} x\, dP_{X|Y}(x|Y = y) \\ &= \int_{-\infty}^{\infty} x\, p_{X|Y}(x|y)\mathrm{d}x \\ &= \int_{-\infty}^{\infty} x\, \frac{p_{XY}(x,y)}{p_Y(y)}\mathrm{d}x,\ p_Y(y) > 0. \end{aligned} \tag{A.84}$$

Therefore,

$$
\begin{aligned}
E[E[X|Y]] &= \int_{-\infty}^{\infty} E[X|Y=y]dP_Y(y) \\
&= \int_{-\infty}^{\infty} \left(\int_{-\infty}^{\infty} x \, \frac{p_{XY}(x,y)}{p_Y(y)} dx \right) p_Y(y)dy \\
&= \int_{-\infty}^{\infty} x \left(\int_{-\infty}^{\infty} p_{XY}(x,y)dy \right) dx \\
&= \int_{-\infty}^{\infty} xp_X(x)dx \\
&= E[X].
\end{aligned}
\tag{A.85}
$$

Example: the expected value of the product $X \cdot Y$, in terms of the conditional expectancy of X, given that Y has occurred, can be calculated from the previous result.

$$
E[XY|Y=y] = E[Xy|Y=y] = yE[X|Y=y],
$$

therefore,

$$
E[XY|Y] = YE[X|Y].
$$

Calculating the expected value yields

$$
E[XY] = E[YE[X|Y]]. \blacksquare
$$

For the random variables X and Y, defined in the probability space $(\mathbb{R}, \mathcal{B}(\mathbb{R}), P)$, the conditional variance is given by

$$
\begin{aligned}
V[X|Y=y] &= E[(X - E[X|Y=y])^2] \\
&= E[X^2|Y=y] - E^2[X|Y=y],
\end{aligned}
\tag{A.86}
$$

which represents the expected value of the deviations of the variable in relation to the conditional expectancy.

For the random variables X, Y and Z, defined in the probability space $(\mathbb{R}, \mathcal{B}(\mathbb{R}), P)$, the conditional covariance is given by

$$
C[X,Y|Z=z] = E[XY|Z=z] - E[X|Z=z] \cdot E[Y|Z=z], \tag{A.87}
$$

considering the existence of the statistical expectancies.

A.18 Two-Dimensional Characteristic Function

The two-dimensional characteristic function is defined as the two-dimensional Fourier transform of the joint density $p_{XY}(x, y)$. It is particularly appropriate for the calculation of moments and the sum of random variables, and is commonly used to demonstrate the Central Limit Theorem.

The definition of the characteristic function in two dimensions is the generalization of the unidimensional function

$$P_{XY}(\omega, \nu) = \mathrm{E}[e^{-j\omega X - j\nu Y}], \tag{A.88}$$

that is,

$$P_{XY}(\omega, \nu) = \int_{-\infty}^{\infty} \int_{-\infty}^{\infty} p_{XY}(x, y) e^{-j\omega x - j\nu y} \, dx \, dy. \tag{A.89}$$

Therefore, the pdf can be obtained with the inverse transform

$$p_{XY}(x, y) = \frac{1}{4\pi^2} \int_{-\infty}^{\infty} \int_{-\infty}^{\infty} P_{XY}(\omega, \nu) e^{j\omega x + j\nu y} \, d\omega \, d\nu. \tag{A.90}$$

If the random variables are independent,

$$P_{XY}(\omega, \nu) = \mathrm{E}[e^{-j\omega X}] \cdot \mathrm{E}[e^{-j\nu Y}] = P_X(\omega) \cdot P_Y(\nu). \tag{A.91}$$

For a sum of N independent random variables $Y = X_1 + X_2 + \cdots + X_N$, one may write

$$P_Y(\omega) = \prod_{k=1}^{N} P_{X_k}(\omega). \tag{A.92}$$

And, in case they are identically distributed, then

$$P_Y(\omega) = P_X^N(\omega). \tag{A.93}$$

Applying the inverse Fourier transform, results in a pdf that can be written as the convolution of the original functions

$$p_Y(y) = \underbrace{p_X(x) * \cdots * p_X(x)}_{N \text{times}}. \tag{A.94}$$

Example: an important application for the property of the sum of independent random variables is the power-flow in power systems analysis. Usually, the companies determine maximum, mean, and minimum values for residential consumption, and later compute the power-flow in the network for the three cases, and this has an elevated computational cost.

However, the energy consumption is evidently random, and it does not make sense to consider that all users have the same electricity usage. Suppose that the energy consumption for each house, X_k, is characterized by a probability distribution $p_{X_k}(x)$, with mean value m_{X_k} and standard deviation σ_{X_k}.

The total load-flow of a city with N houses is modeled by the distribution of the sum of all the individual load-flows, $Y = X_1 + X_2 + \cdots + X_N$, that is,

$$P_Y(\omega) = \prod_{k=1}^{N} P_{X_k}(\omega),$$

in which $P_{X_k}(\omega)$ represents the characteristic function of each consumer. The calculation can be quickly done using the Fast Fourier Transform (FFT) algorithm. ■

After obtaining the characteristic function of the total load-flow, it is sufficient to calculate the Inverse Fourier Transform (IFT), using the same algorithm, to obtain the pdf that represents the energy consumption for that network. From this distribution, it is possible to calculate all the moments, including the mean and standard deviation, to obtain a statistical estimate of the total consumption.

The bi-dimensional characteristic function, also called moment generating function, can be used to calculate the mean, standard deviation, and all the moments of the random variables X and Y,

$$\mathrm{E}[X^n Y^m] = \frac{1}{(-j)^{n+m}} \cdot \frac{\partial^{n+m}}{\partial \omega^n \partial \nu^m} P_{XY}(\omega, \nu) \Big|_{\omega=0, \nu=0}. \tag{A.95}$$

A.19 Sum of Random Variables

The characteristic function is also used to determine the pdf of a random variable that is the sum of other independent random variables, $Z = X + Y$.

For this particular case, the characteristic function of the resulting random variable is

$$P_Z(\omega) = \mathrm{E}[e^{-j\omega Z}] = \mathrm{E}[e^{-j\omega(X+Y)}] = P_X(\omega) \cdot P_Y(\omega). \tag{A.96}$$

It is possible to write,

$$p_Z(z) = \int_{-\infty}^{\infty} p_X(\rho)p_Y(z - \rho)d\rho, \qquad (A.97)$$

or, in an equivalent form,

$$p_Z(z) = \int_{-\infty}^{\infty} p_X(z - \rho)p_Y(\rho)d\rho. \qquad (A.98)$$

Therefore, the sum of independent random variables results in the convolution of their respective probability density functions. This is a powerful property, usually stated as a theorem, that can be used to prove the Central Limit Theorem, and also to determine the bit, or symbol, error probability for a channel with additive noise.

A.20 Function of Joint Random Variables

The transformation of joint random variables, also called, the function of joint random variables can be used to solve problems in several areas of Mathematics, Physics, Engineering, Economics, to mention only a few. It is also useful in Electronics, Communications, Control, and Power Systems.

In order to obtain the formula that maps the input random variables (X, Y) into the output random variables (U, V), it is necessary to take into account that the probability measure is preserved for any transformation, as depicted in Figure A.44.

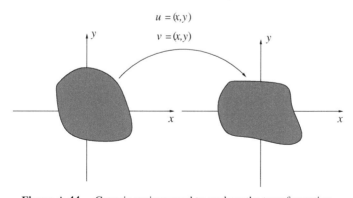

Figure A.44   Generic regions used to analyze the transformation.

That is, the differential volumes at the input and output are equal

$$p_{XY}(x, y)dxdy = p_{UV}(u, v)dudv, \qquad (A.99)$$

in which $dE = dxdy$ and $dS = dudv$ represent, respectively, the differential areas that are spanned by the joint input and output variables.

In Figure A.45 the square represents the differential area produced by the input random variables. The vertices of the Borel rectangle are the coordinates $A = x, y + dy$, $B = x + dx, y + dy$, $C = x + dx, y$ and $D = x, y + dy$.

At the output, the joint random variable is transformed by the respective functions, $U = f(X, Y)$ and $V = g(X, Y)$. This mapping produces a new differential area, which is shown in Figure A.46, whose vertices are

$$E = (f(x, y + dy), g(x, y + dy)),$$

$$F = (f(x + dx, y + dy), g(x + dx, y + dy)),$$

$$G = (f(x + dx, y), g(x + dx, y)),$$

$$H = (f(x, y), g(x, y)).$$

Because the area is infinitesimal, it can be approximated by the parallelogram illustrated in Figure A.47, and the vertices, obtained by the linearization process, are

$$E = \left(u + \frac{\partial f(x, y)}{\partial y}dy, v + \frac{\partial g(x, y)}{\partial y}dy\right),$$

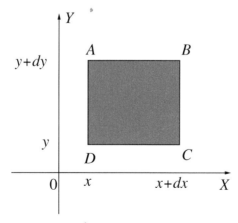

Figure A.45   Region $\{x < X \leq x + dx, \ y < Y \leq y + dy\}$.

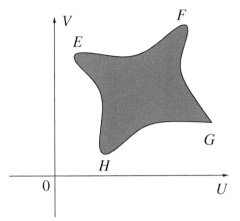

Figure A.46 Differential region for the output joint variables.

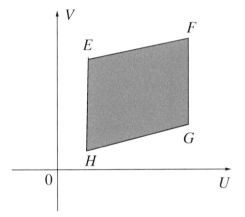

Figure A.47 Linearized output differential region.

$$F = \left(u + \frac{\partial f(x,y)}{\partial x} dx + \frac{\partial f(x,y)}{\partial y} dy, v + \frac{\partial g(x,y)}{\partial x} dx, + \frac{\partial g(x,y)}{\partial y} dy \right),$$

$$G = \left(u + \frac{\partial f(x,y)}{\partial x} dx, v + \frac{\partial g(x,y)}{\partial x} dx \right),$$

$$H = (u, v).$$

The area of the parallelogram, with vertices $A = (x_4, y_4)$, $B = (x_3, y_3)$, $C = (x_2, y_2)$, $D = (x_1, y_1)$, is given by the formula (Bronstein and Semendiaev, 1979)

$$S = (x_1 - x_2)(y_1 + y_2) + (x_2 - x_3)(y_2 + y_3) + (x_3 - x_1)(y_3 + y_1). \quad \text{(A.100)}$$

Therefore, one obtains

$$dS = \left(u - u - \frac{\partial f(x, y)}{\partial x} dx \right) \left(v + v + \frac{\partial g(x, y)}{\partial x} dx \right)$$

$$+ \left(u + \frac{\partial f(x, y)}{\partial x} dx - u - \frac{\partial f(x, y)}{\partial x} dx - \frac{\partial f(x, y)}{\partial y} dy \right)$$

$$\cdot \left(v + \frac{\partial g(x, y)}{\partial x} dx + v + \frac{\partial g(x, y)}{\partial y} dy \right)$$

$$+ \left(u - u + \frac{\partial f(x, y)}{\partial x} dx + \frac{\partial f(x, y)}{\partial y} dy \right)$$

$$\cdot \left(v + v + \frac{\partial g(x, y)}{\partial x} dx + \frac{\partial g(x, y)}{\partial y} dy \right).$$

This result can be simplified to

$$dS = \left(-\frac{\partial f(x, y)}{\partial x} dx \right) \left(2v + \frac{\partial g(x, y)}{\partial x} dx \right)$$

$$+ \left(-\frac{\partial f(x, y)}{\partial y} dy \right) \left(2v + \frac{\partial g(x, y)}{\partial x} dx + \frac{\partial g(x, y)}{\partial y} dy \right)$$

$$+ \left(\frac{\partial f(x, y)}{\partial x} dx + \frac{\partial f(x, y)}{\partial y} dy \right) \left(2v + \frac{\partial g(x, y)}{\partial x} dx + \frac{\partial g(x, y)}{\partial y} dy \right).$$

Multiplying the terms and simplifying the result, gives

$$dS = \left| \frac{\partial f(x, y)}{\partial x} \frac{\partial g(x, y)}{\partial y} - \frac{\partial g(x, y)}{\partial x} \frac{\partial f(x, y)}{\partial y} \right| dx dy.$$

The last expression represents the Jacobian of the transformation and can be written using the determinant notation

$$J(x, y) = \left| \begin{matrix} \partial u / \partial x & \partial u / \partial y \\ \partial v / \partial x & \partial v / \partial y \end{matrix} \right|. \quad \text{(A.101)}$$

Substituting the result into Equation A.99, yields

$$p_{UV}(u, v) = \frac{p_{XY}(x, y)}{|J(x, y)|}, \quad \text{(A.102)}$$

which is a compact formula to compute the pdf of the output joint random variables, given the input and the Jacobian.

Example: consider the transformation obtained by the equations $U = aX + bY$ and $V = cX + dY$.

The Jacobian is

$$J(x, y) = \begin{vmatrix} a & b \\ c & d \end{vmatrix} = ad - cb$$

and the inverse functions are given by

$$x = \frac{1}{a}(u - by),$$

$$v = \frac{c}{a}(u - by) + dy \qquad \frac{a}{c}v = u - by + \frac{da}{c}y,$$

$$y\left(\frac{da}{c} - b\right) = \frac{a}{c}v - u \Rightarrow y = \left(\frac{a}{c}v - u\right)\frac{c}{da - cb},$$

$$y = \frac{uc - av}{cb - da},$$

$$x = \frac{1}{a}\left[u - \left(\frac{uc - av}{cb - da}\right)\right] = \frac{du - bv}{ad - bc}.$$

Therefore, the joint pdf at the output is given by

$$p_{UV}(u, v) = \frac{p_{XY}\left(\frac{du-bv}{ad-bc}; \frac{uc-av}{cb-bc}\right)}{|ad - cb|}. \blacksquare$$

Example: it is possible to solve $U = X + Y$ using an auxiliary variable $V = X$.

Using the previous formula,

$$p_{UV}(u, v) = \frac{p_{XY}(x, y)}{|J(x, y)|},$$

in which

$$p_U(u) = \int_{-\infty}^{\infty} p_{UV}(u, v)\, dv.$$

But,

$$J(x, y) = \begin{vmatrix} 1 & 1 \\ 1 & 0 \end{vmatrix} = -1,$$

then

$$p_{UV}(u, v) = \frac{p_{XY}(x, y)}{1} \qquad \text{in which}: \ x = v, \ y = u - v.$$

Therefore,

$$p_{UV}(u, v) = p_{XY}(v, u - v)$$

and

$$p_U(u) = \int_{-\infty}^{\infty} p_{XY}(v, u - v)\, dv. \blacksquare$$

If X and Y are independent random variables, the previous equation simplifies to

$$p_{XY}(x, y) = p_X(x) \cdot p_Y(y) \qquad \Rightarrow \qquad p_U(u) = \int_{-\infty}^{\infty} p_X(v) \cdot p_Y(u - v)\, dv.$$

This result indicates that the sum of random variables produces the convolution of the respective probability density functions, as previously established.

B

Glossary of Economics

"We publish so as not to spend our lives correcting drafts. I mean, we publish a book to get rid of it."
Jorge Luis Borges

This appendix presents the main economic definitions that are useful to understand the subject (Mankiw, 2010) (Smith, 1986) (Siegel and Shim, 1987) (Blomqvist et al., 1987) (Chacholiades, 1986) (Kurtz, 1975) (Fischer, 1898) (Cournot, 1897) (Camacho and da Silva, 2018) (de Azevedo and cei ção, 2017) (Koutsoyannis, 1977) (Niall Ksihtainy, 2013) (Balckburn, 1997) (Heilbroner and Thurow, 1982) (Economia.net, 2020) (Estadão, 2020) (World Economic Forum, 2018) (Alvaredo et al., 2018) (Roser, 2013) (Comin, 2020) (Alves, 2001) (Ventura, 2018) (Frazão et al., 2020) (InfoMoney, 2020) (Folha de São Paulo, 2020) (Editora Melhoramentos Ltda., 2020) (BTG Pactual Digital, 2020).

Accounting Profit – The amount of resources remaining for the owners of a firm, after compensation for all factors of production, except capital.

Acquisition – Purchase of one company by another, in which both maintain their respective legal identities, in contrast to what occurs in a merger or incorporation. In the case of corporations, it may only mean the acquisition of a lot of shares large enough to allow the company to take control of the company.

Actual Investment – Investment including unwanted inventory accumulation.

Actuary – Professional who uses mathematical or statistical methods to determine the loss ratio, in addition to developing systems for calculating future premiums.

Default – Compliance with a contractual obligation.

ADR – The American Depository Receipt is a receipt issued by a North American depository bank, which represents shares of a foreign issuer that are deposited and in the custody of this bank.

AFC – Advances on foreign exchange contracts, which represent a form of financing used by exporting companies, through which the company receives resources in advance, which are used to finance its production.

AFD – Advance on foreign exchange delivered, equivalent to ACC, with the difference that the capital advance occurs when the goods are ready and shipped, and can be requested within 60 days after shipment.

Aggregate – Corresponds to the total of a country's entire economy.

Aggregate Demand – Annual sum of investments, net exports, and consumption expenditures of the population and government in a country.

Aggregate Offer – Total quantity of goods and services that would be offered for sale at various possible average prices.

Aggregate Savings – Gross savings, variation before discounting the amortizations suffered by the equity value over the period considered.

Agreement – Administrative act that allows the release of the resource provided for in the General Budget of the Union (GBU). When the government intends, for example, to finance a project in a municipality, it signs an agreement with the mayor.

Amortization – Payment of debt contracted by companies, or by the government, with financial institutions in the international market.

Annual Allowance – Additional salary, calculated based on the value of a minimum wage paid annually to workers.

Anticipation of Budget Revenue – Short-term loan operation granted by banks to the federal, state, or municipal government with the guarantee of expected revenues.

Anti-cyclical – What is meant to prevent, overcome, or minimize the effects of the economic cycle.

Aperture – Generic term used to define the price of the first trade of the day for a particular asset listed on the stock exchange, which can be a stock or a futures contract, for example.

Application – Operation in which the buyer, an investor, or a company, acquires part or all of the shares of a company.

ARPU – Average Revenue Per Use, which means average revenue per user, being a term widely used to measure the performance of telecommunications operators.

Assets – Set of assets and credits that make up an economic subject's assets.

Asset Turnover – Financial analysis indicator that indicates the efficiency with which the company uses its assets to generate sales. calculated as the division of net sales revenue by the company's total assets. The higher the index, the greater the company's efficiency in the use of its assets.

Audit – Confirmation of accounting records and statements, obtained by examining all the documents, books, and records of a company or entity.

Authorized Capital – Statutory limit on the powers of the general meeting or of the board of directors to increase a company's share capital.

Automatic Zeroing – Operation whereby the Central Bank purchases government securities from banks by committing these institutions to repurchase such securities on the day after trading.

Available Values – Set of the company's liquidity or credit securities that can be quickly converted into currency.

Average Profitability – Average of the percentage of income obtained on a financial investment during a given period.

Average Propensity for Consumption – The ratio between consumption and income (APC).

Bachelor of Economic Sciences – Is the professional who performs functions in the areas of economics and finance, with higher education in Economic Science.

Balance of Current Transactions – Balance of trade balance, exports less imports, and services such as payment of interest on foreign debt.

Balance of Payments – Records the result of all transactions, such as goods, services, transfers, and capital flows, between a country and the rest of the world

Balance of Trade – Records all exports and imports made by companies in a country.

Balance on Current Transactions – Result of all the country's operations abroad, including income and expenses from the trade balance (exports and imports), from the service account (interest, international travel, transportation, insurance, profits) dividends, miscellaneous services) and unilateral transfers.

Balance Sheet – Financial statement that details and quantifies the assets, liabilities and equity of a company.

Balance Sheet – Financial statement that lists all the assets and liabilities that a company has in a given period.

Ballast – Gold deposit that serves as a guarantee for paper money. In the operations of the national financial market, the ballast represents the securities given in guarantee of an open market operation.

Banco Medici – One of the first financial institutions based on international trade was founded in Florence, Italy, in 1397.

Bank Deposit – Sum of money placed under the custody of a banking institution.

Banking System – Set of private and state financial institutions that offer services such as custody, loan, and investment of money.

Bank – Institution authorized to store money, securities, and securities of third parties and dispose of them for loans, also performing related operations, such as collections, payments, transactions with foreign currencies, among others.

Bankruptcy – Situation in which a company can no longer afford to pay its liabilities, or when the company's liabilities exceed the fair value of its assets. Thus, a company in bankruptcy has negative equity.

Base – Sum of currency issued and bank reserves held by a country.

Basic Interest Rate – Annual interest rate fixed by a bank, which serves as a reference for calculating the different conditions offered by this bank.

Benchmarking – Defines the process used to evaluate the performance of a financial asset in relation to others identified as having the best performance in the sector, or category, of investment.

Bias or Systematic Error – Systematic distortion between the measurement of a statistical variable and the real value of the quantity to be estimated. The introduction of a bias in the statistical calculation may be linked to the imperfection or deformation of the sample that serves as the basis for the estimate, or to the evaluation method itself.

Bid – Price offered by representatives of brokerage firms in a public auction, for the purchase or sale of a lot of shares.

BID – Inter-American Development Bank. International institution, based in Washington, USA, to financially assist the development of infrastructure in emerging countries.

Billing – Total value calculated from sales of a company's products or services.

Bill of Exchange – Funding instrument used by credit, financing, and investment companies, which is issued based on a commercial transaction. Commercial title, by means of which a creditor, called an issuer, orders the debtor, or drawer, to pay, within the indicated period, a precise amount to a designated third person, the beneficiary. The bill of exchange, which became the standard of payment in European commerce, was created around 1400.

BIS – The Bank of International Settlements was created in 1930 to promote cooperation between European central banks and to provide additional facilities for international financial transactions.

Bitcoin – Encrypted digital currency that allows financial transactions free from institutions, but monitored by network users and encrypted in a decentralized database (blockchain). Bitcoin does not derive from physical currencies, has no financial backing, and is not recognized by the Securities and Exchange Commission (SEC) as an asset.

Black Market – Part of the active economy that involves goods or services banned in a given region, for example, trade in unregistered weapons, trade in illicit drugs or counterfeit products.

Blockchain – Decentralized database that records financial transactions, which are stored on computers around the world. The database records the sending and receiving of digital currency values in cryptographic format, and the parties must authorize access to each other.

Blue Chip – Stock with great liquidity and tradition, with high demand in the stock market by traditional and large investors.

BNDES – Brazilian National Bank for Economic and Social Development. A company wholly owned by the Brazilian government, responsible for implementing its long-term credit policy.

Bonus – Shares distributed free of charge to shareholders, as a result of a capital increase made through the incorporation of reserves. Similar to promissory notes, bonuses are fixed income bonds issued by companies, banks, or governments.

Book Entry Share – A security that circulates in the capital markets without the issuance of certificates or precautions, being registered by a bank, which acts as a depository for the company's shares and processes payments and transfers by issuing statements bank accounts.

Book Value – Sum of the company's market value plus its net debt. Indicator both the value of the company belonging to the shareholders, which is called market value, and that which belongs to the creditors, which is the net debt.

Bradies – Foreign debt securities of developing countries, renegotiated according to the rules of the Brady Plan, prepared by the former US Treasury Secretary, Nicholas Brady. The plan aimed at issuing bonds to replace the foreign debt of these countries and, in the conversion, a discount was applied to the value of loans.

Break Even Point – Balance between revenue and expense. From the break even point the company's revenue generates, proportionally, a higher profit.

Breakeven Point – Point at which income and expense match. From this point on, the company's revenue generates, proportionally, a higher profit.

Brokerage – Rate of remuneration of a financial intermediary in the purchase or sale of securities.

Budget allocation – The amount determined in the general budget, to meet each expense.

Budget – Limiting forecast of the monetary amounts that must be used as expenses and income, over a given period, by an individual or by a company.

Budget Restriction – The limit that income places on spending.

Business Plan – Document containing details of products or services, markets, future strategy, and resumés of the company's main executives.

Cable Dollar – Quotation for the purchase or sale of the US currency outside the conversion channels authorized by the Central Bank.

Call for Capital – Determines a subscription for new shares, with or without premium, to increase the capital of a Company.

Capex – Abbreviation for capital expenditure, and defines the investments made in capital goods that aim at the continuity or expansion of a company's operations.

Capital Balance Sheet – Accounting for all movements related to assets and liabilities, which occurred between a country and the rest of the world, in a given period.

Capital centralization – Capital concentration process, which necessarily implies the elimination of autonomous blocks of capital that start to operate in a centralized manner.

Capital Concentration – Increase in the size of a company or block of capital due to the accumulation process.

Capital Expense – Defines the expenses incurred by a company in purchasing an asset.

Closed Capital – A company that operates as a limited liability company and has its shares in the hands of a few shareholders, and these securities are not traded on the stock exchange.

Capital Flow – Movement of capital between countries, which enables international investments.

Capital Gain – Difference between the income received from the sale of a certain asset, such as shares or real estate, and the cost of acquiring this asset.

Capital Goods – Goods that are used for the production of other goods, such as machines and equipment.

Capital Growth – Absolute concentration of capital, that is, a real increase in the capital of a company or group.

Capital Increase – Terminology used to reflect changes in a company's capital structure, through the incorporation of new resources, or reserves, to its capital.

Capitalism – Economic system based on private ownership of the means of production and their use for profit. Being a regime based on the dissociation between the owners of the means of production. These means are used for the purpose of profit. Employees carry out production by paying a wage, which remunerates their workforce.

Capitalist Economy – Economic market system based on private property and initiative.

Capitalist – Individual who temporarily invests in emerging companies with evident growth potential, and whose objective is to obtain profitability above the alternatives available in the financial market. In general, this individual participates in the management of the company for the duration of the investment.

Capitalization Bond – Bond purchased for a low price, for which the buyer competes for sweepstakes and earns interest at the time of redemption.

Capital Markets – Set of institutions that provide long-term capital. Includes stock exchanges, banks, and insurance companies.

Capital Remuneration Rate – The capital remuneration rate (CRR) allows comparing the profitability of the business with alternative investments in the financial market. CRR results from the division of the net margin by the invested capital. The net margin is equal to the gross income minus the direct cost, called the effective operating cost, subtracted from the costs corresponding to the depreciation of improvements and equipment, less the costs related to labor. It is equivalent to the rate of return on capital and the rate of return to capital.

Capital Structure – Nature of the main shareholders, the form of control, as well as the degree of concentration of ownership.

Capital – Sum of all resources, assets, and values, mobilized for the constitution of a company.

Capital to be paid in – Part of the capital subscription that the shareholder of a company must still pay, that is, that has not yet been placed in the company.

Cartel – Agreement between competing companies, with the objective of limiting, or suppressing, the risks of competition, in addition to increasing product prices. It aims to increase profits, to the detriment of consumers, by setting prices or production quotas, dividing customers and markets, or by coordinated action between participants.

Cash Flow – Statement of origin and application of funds disclosed by companies, which is annual.

Cash Flow – The cash flow represents the forecasts and the record of the movement of money in and out of a company, government agency or group of people.

Cash in Kind – Paper or currency held by the population in a country.

Cash Turnover – Indicator calculated as the division between the company's revenues and its working capital.

CBB – Central Bank Bonds. Paper with fixed rates with a minimum term of 28 days. The banks inform the rate they want to buy the paper and the Central Bank evaluates whether it accepts or not, with the objective of making monetary policy.

C-Bonds – Brazilian foreign debt security most traded on the international market.

Central Bank of Brazil – Federal agency created to inspect financial institutions. The Central Bank (CB) is responsible for regulating the government's monetary and credit policies, managing international reserves, and inspecting the National Financial System.

Centralized Economy – Economic system in which the government is responsible for most economic decisions.

Circuit Breaker – Interruption of the trading session, whenever the index representative of the prices of a set of shares has fallen substantially.

Classical Economics – In 1817, David Ricardo laid the foundations for free trade and the specialization of work.

Closed Economy – An economy that does not engage in international trade.

Cold Duplicate – Illegal duplicate, issued without effecting the sale of goods.

Collateral – Financial asset or other type of asset pledged as collateral by a borrower to provide collateral for a given loan.

Collectivization – The Soviet Union began compulsory collectivization of agriculture in 1929 under Josef Stálin.

Commercial Dollar – Exchange rate that is published by the Central Bank and used in the country's trade balance and services operations, such as exports and imports, in the payment of external debt service and in the remittance of dividends by companies with headquarters abroad.

Commercial Exchange – Dollar quotation, used to close export and import contracts. The commercial exchange also registers loan operations by companies abroad, foreign direct investments, inflows, and outflows of foreign investments in fixed income and on stock exchanges.

Commercial Paper – Category of debt securities issued by companies, in the local or international market, whose resources are used, for example, to finance the company's short-term activities or working capital needs, such as purchasing stocks, paying suppliers.

Commitment – Reserve of funds made by the government in the Treasury, before releasing the funds provided for in the General Budget of the Union (GBU).

Commodities and Futures Exchange – Institution where goods are traded, especially the most important in the domestic and international markets, whose stocks can be existing or projected.

Commodities Exchange – Centralized market for transactions with goods, mainly commodities, that is, raw products in the raw state with high commercial importance, both in the domestic market and in the foreign market, including ores, coffee, cereals, and cotton.

Commodities – Primary products, or commodities, such as coffee, soy, corn, wheat, oil, in addition to some semi-finished industrial products such as cellulose and sugar.

Compensation – Account adjustment process between individuals or companies through determining and adjusting the differences between credits and debits.

Compliance – Set of actions to enforce legal and regulatory standards, policies or guidelines established for a business, as well as to avoid any deviations.

Compulsory Deposit – Deposit, made at the Central Bank (BC), of a percentage of the sums deposited in commercial banks.

Conglomerate – Company or economic group that operates in several sectors of the economy, without necessarily following criteria of technical, productive or commercial complementarity.

Conjuncture – Elements constituting the economic situation of a sector, industry, region or country at any given time.

Consortium – A system through which a closed group of individuals, whether natural or legal, unites in order to create savings that allow the purchase of durable movable goods, real estate, and tourist services, through self-financing.

Consumer Goods – These are those used directly by the general public. The rest are capital goods.

Consumer Price Index – The Consumer Price Index (CPI) is a measure of the weighted price of a basket of typically purchased goods and services in the economy.

Consumption Bubble – Increase in consumption observed during a given period.

Consumption – Purchases of goods and services by buyers.

Control Group – Individual, group of individuals, with or without family ties, legal entities under private or public law, who ultimately exercise shareholder control over a company or economic group.

Controlling Shareholder – Designates the individual, company, or group of people who, through their rights as a partner, are able to control the voting and deliberations during the company's general meeting, as well as appoint the majority of the directors of the company. company.

Cooperative – Company made up of people, who have similar marketing interests.

Corporation – An association of shareholders with a statute that guarantees certain legal powers, privileges, and responsibilities, separate from those of individual owner shareholders.

Corporatism – A system that provides leadership and regulatory power to single unions, employers, or workers, constituted by profession.

Costing – Set of government daytime expenses. The necessary expenses for the ministries, departments, and agencies of the public administration to function. This type of expense appears as other current expenses and does not include payroll expenses.

Cost of Capital – The cost of capital of a company can be defined as the rate of return that the company must obtain on its investments to keep its market value unchanged.

Cost – Sacrifice, measured by the price paid to purchase, produce or maintain goods or services. It also refers to the valuation of an acquired good or service.

Coupon – Rate determined when a fixed income security is issued, for which an issuer undertakes to pay interest at periodic intervals.

Crash – Denomination given to a sharp drop in the stock exchanges, typically associated with financial disaster.

Credit Market – Set of institutions, banking and non-banking, which provides short-term capital, both for consumption and for working capital of companies, in addition to operating in the public securities market.

Cryptocurrency – A type of currency, or virtual asset, that uses cryptography to ensure more security in financial transactions on the Internet. There are several types of cryptocurrency, Bitcoin being the best known of them.

Cryptography – Transformation of readable data into a form that cannot be understood, in order to protect information.

Cryptology – Study of secret codes, or ciphers, and the devices for creating and deciphering those codes.

Currency Appreciation – An increase in the price of a currency, compared to another currency, or other currencies, in a flexible exchange rate system.

Currency Correction – Change in national currency prices in relation to foreign currencies. Also known as an exchange rate adjustment.

Currency Depreciation – A reduction in the price of a currency, compared to another currency, or other currencies, in a flexible exchange rate system.

Current Account Deficit – It is the result of commercial transactions in a country like the world, including exports and imports, plus services, so-called unilateral transfers, insurance costs, sea freight, and other foreign trade expenditures.

Current Assets – Sum of all the assets of a company that can, in the short-term, up to one year, be converted into liquidity, that is, sold in order to increase the company's cash.

Current Liquidity – Determines how much this company has to receive in the short-term in relation to each currency unit that must pay in the same period.

Custody Agent – Financial institution responsible for managing its own accounts and customer accounts, with custody services, and custody accounts can be maintained in the name of your customers.

Custody – Securities and securities custody service provided to investors

Day Trade – It is the purchase and sale of shares or bonds on the same day. In stock trading, trades labeled day trade can only be carried out in the futures and options markets, and are prohibited in the spot market.

Dealer – Financial institution that has a direct line with the Central Bank and, therefore, the *dealer* must transmit to the financial market the communications of the Central Bank on auctions, interest, among other obligations, such as bidding on all securities auctions.

Debentures – Securities that represent a loan to a limited company, which yields interest and monetary restatement, are guaranteed by the company's assets and have a preference for redemption over almost all other debts.

Debt Structure – Level of use of external funds, which imply financial liabilities. It also refers to the degree of dependence in relation to the banking system.

Deductible – Part of the indemnity that is the responsibility of the insured, that is, how much you have to pay to be entitled to receive the indemnity.

Default – Breach of any clause of a contract related to creditor and debtor, known as default in the jargon of the financial market.

Defensive Buying – Preventive purchase, made by a company, of its own shares, either to ensure control against possible rivals, to create room for maneuver for future merger operations, or to reduce the control threshold.

Deficit – Negative result when expenses are subtracted from revenue.

Deflation – Negative change in prices of the economy. Indicates decreasing production and consumption of goods and services produced in the country.

Demand Deposit – Bank deposit, with immediate funding and with no income to the account holder, which allows free movement of money.

Demand – Quantity of a good or service that can be purchased for a defined price, in a given market, during a specific unit of time.

Depreciation – Defines a debt that aims to reduce the accounting value of a given asset. This launch seeks to account for the loss of value of an asset due to use, time, technological obsolescence, or reduction in the market price.

Depression – Phase of the economic cycle, characteristic of capitalist economies, marked by a decrease in production, a tendency to lower prices and an increase in unemployment. The crash of the Wall Street stock exchange in 1929 represented a drastic drop in the heat of the securities and stocks traded and started the Great Depression in the United States.

Derivative – Financial instrument whose characteristics are linked to other securities, or assets, that serve as a reference. Example: stock options, futures contracts on the commercial dollar.

Direct Investment – Long-term foreign investment aimed at equity participation in existing companies or the installation of new units.

Discount – Amount deducted from the face value of promissory notes, bills of exchange and trade bills, when they are paid before the stipulated deadline.

Discount – Difference between the market value and the nominal value of a security. If the market value or amount paid is less than the nominal value, the difference is called negative goodwill. If it is greater than the nominal value, the difference is called goodwill. Discount on the price of a security, which generally includes interest and monetary restatement, or currency restatement.

Diversification – Growth process, which implies an expansion of the performance of the company or group with an increase in the diversity of activities, whether in terms of the market or the technological base.

Dividend – Portion of a company's profits, distributed to shareholders as a form of remuneration.

Domestic Production – Set of goods and services produced by the national economy, in the national territory, whatever the nationality of the producers.

Dow Jones – New York Stock Exchange Index, created in 1986. It represents the profitability of the group of stocks most frequently traded daily on the trading floor.

Downturn – Process by which the aggregate demand of a country's economy retracts.

Draw Back – Tax exemption for imports made in order to industrialize the product in the country and later export it.

Dry Liquidity – Reflects a company's ability to meet its short-term obligations, with the difference that inventories are excluded from the company's current assets.

Dumping – Unfair pricing practices, which are practiced below the cost in international trade, or at the cost of production, with the objective of eliminating competitors and gaining larger market shares.

Duopoly – Particular case of an oligopoly in which there are only two sellers of a certain commodity or service.

Durable Consumer Goods – Consumer goods that provide service for a long period of time, as is the case, for example, with washing machines, refrigerators and automobiles.

EAP – The Economically Active Population is the contingent of workers with a formal contract, being the basis for calculating the unemployment rate.

Earnings – Denotes benefits, such as dividends, bonuses, subscription rights, interest on capital, and others, distributed by a company to its shareholders.

EBITDA – Earnings Before Interest, Taxes, Depreciation and Amortization.

ECLAC – Economic Commission for Latin America, a regional institution of the United Nations that aims to study and create development possibilities for Latin American countries.

Econometrics – Strand of Economics that uses a set of statistical tools, mainly estimation and hypothesis testing, to model and understand the relationships between economic variables.

Economic Activity – Set of procedures by which people satisfy their needs, through the production and exchange of goods and services.

Economic Agent – Individuals, groups of individuals or bodies that constitute, from the point of view of economic movements, the centers of decision and fundamental actions.

Economically Active Population – Set of people who are inserted in the labor market, according to the Brazilian Institute of Geography and Statistics (IBGE).

Economic Analysis – Application to the economic reality of the scientific method of decomposing a problem into a set of elements that are easier to understand.

Economic Block – Set of countries, located in the same region, that establish a common economic and commercial policy.

Economic Development – Process, or set of economic activities, by which the Gross Domestic Product (GDP) increases and the consequent improvement in the population's standard of living.

Economic Group – Group of companies subordinate to a single decision center that, through financial, personal, and ownership links, is able to exercise power, at least in strategic terms.

Economic Growth – Growth in the volume of goods and services produced in a country, as measured by the evolution of the GDP.

Economic Indicator – Quantitative instrument that measures a quantity or an economic phenomenon.

Economic Instability – Economic situation in which fluctuations are observed in the level of production, employment, or consumption.

Economic Opening – Government policy for the gradual abolition of restrictions on flows with foreign countries and international capital movements.

Economic Policy – Also known as Normative Economics, it represents the set of practical government actions with the purpose of conditioning, guiding and guiding the economic system, in order to achieve certain politically established economic objectives.

Economic Regime – Institutional characteristics that determine, in the scope of an economic system, the conditions for the practical organization of the functioning of the economy.

Economic Slowdown – Process by which the supply of jobs, production, and demand for national products decreases.

Economics – Science that studies the economic laws that must be followed in order to maintain a high level of productivity, to improve the life standard the populations, with the correct use of resources.

Economic Stability – Indicates the maintenance of full employment, general price stability, and balance of the international balance of payments.

Economic System – Set of related legal and social institutions, in which certain technical means are employed, organized according to certain dominant causes, to ensure the realization of the economic balance.

Economist – Professional who performs the functions of economics and finance, with higher education in Economic Sciences and registration at the respective order.

Effective Shareholder – Final owner of shares in a company, which may or may not be represented by a nominal shareholder.

Elastic Demand – Demand that has greater elasticity than the unit. A drop in price causes an increase in total spending on the product in question.

Elastic Offer – Offer that has greater elasticity than the unit.

Endorsable Share – Title that can be transferred by endorsement on the back of the note.

Endorsement – Act whereby a third person, other than the drawer, endorsers or guarantees the payment of a security on the due date.

Endorsement – Assignment, by the bearer of a commercial title to his order, of his signature on the back of the title, to transfer the represented credit to a third party.

Equity – Common share. Excess of assets over liabilities, that is, the net worth of a company. It is also the share capital plus the surplus in a corporation.

Equity Fund – Fund in which the funds raised from the shareholders are invested in a diversified portfolio of shares, with the result of the income from this portfolio being distributed to the shareholders, in proportion to the number of shares held, after the deduction of taxes and fees charged by the portfolio administrator.

European Community – Also referred to as the Common Market, the European Community (EC) is a customs union and has other features of economic integration.

Exchange Adjustment – Changes in national currency prices in relation to currencies foreign. Also known as currency correction.

Exchange Devaluation – Increase in prices of foreign currencies in national currency.

Exchange Fluctuation – Procedure by which the exchange rate can vary freely, obeying the laws of supply and demand.

Exchange – Operation in which national currency is exchanged for foreign currency, or vice versa.

Exchange Overvaluation – Excessive reduction in the prices of foreign currencies in national currency.

Exchange Rate Lack – Exchange rate difference caused by inflation and calculated in relation to a currency established as a reference.

Expansion – Situation in which production and demand volumes in the economy show growth rates.

External Accounts – Account system that records the economic transactions of residents of a country with abroad.

External Debt – Process by which a country assumes debts with supranational entities or from other countries. Total debts with government external creditors, including states, municipalities, and state-owned companies.

External Fund – Resources obtained by a company for its current or investment activities, through loans in the banking system, launching of debentures or sale of shares, for example.

Federal Reserve – Central Bank of the United States. The action of the FED regulates interest rates in the American market.

Fictitious Accumulation – Valuation of property titles, shares, or other financial assets, independently of the assets it represents, that is, unrelated to the effective production of wealth. Typical of a speculative process, especially on stock exchanges or a privileged association with the public fund, via the financial market or government agencies.

Fiduciary Currency – Non-backed currency with no intrinsic value, the price of which is determined by the government.

Fiduciary Property – Transfer of ownership of securities to the custodian institution, so that it can exercise all the rights inherent in those securities for the purposes of custody, conservation, the exercise of rights; not being able to freely dispose of them, as it is obliged to return to the depositor, when requested, the amount delivered to it with the changes resulting from the changes in the share capital or number of shares of the issuer.

Financial Agent – Financial institution that can represent, as guarantor, financier or endorser, a public entity.

Financial Allowance – Payment received by the worker for the sale of a third of your vacation for the company.

Financial Analysis – Methodology based on the analysis of financial statements, such as the balance sheet, income statement, and demonstration of a company's origins and resources, with the objective of determining its current financial position and thus project its possible future performance.

Financial Cost – Sum of expenses related to interest, corrections, and fees charged on loans and financing.

Financial Crisis – Situation that reflects an imbalance between an entity's income and expenditure.

Financial Dominance – Structural attribute of inter-capitalist competition, especially within the scope of big capital, which arises from the emergence of financial capital and the resulting empowering of capital at interest.

Financial Engineering – Economic study that aims to structure financial operations that produce high profitability, both for the investor and for the institution that offers them.

Financial Function – Results from the combination of financial management and financial strategy.

Financial Hegemony – Set of processes that allow agents, who have large amounts of capital, to control or influence the decisions of other agents.

Financial Institution – Institution that carries out financial operations such as savings collection, receipt of securities, credit operations, among others.

Financial Intermediation – Process by which financial transactions are carried out through institutions that act as intermediaries between economic agents.

Financial Investment – Financial transaction by which an individual or an entity temporarily transfers the exercise of certain purchasing power to a financial institution in exchange for obtaining returns on the invested capital.

Financial Investments – Transfer of funds to other National Budget accounts or funds, such as constitutional transfers, funds from states and municipalities, or those made to state-owned companies in which the government holds the majority of the share capital.

Financial Leasing – Operation of leasing in which the sum of total leases and other payments provided for in the contract are sufficient for the lessor to recover the cost of the leased asset and obtain a return on invested resources.

Financial Management – Set of routines of a firm or group for the application of cash flow in the portfolio composition, such as share placements, speculation with stocks or currencies, purchase and sale of financial assets.

Financial Mathematics – Mathematics branch that studies capital equivalence over time. Your knowledge is essential to understand and operate in the financial and capital markets, in addition to working in financial management.

Financial Profile – Situation of a firm that, at each moment, is the result of its financial strategy, in terms of its capital structure, debt structure and diversification.

Financial Statement – Set of reports that categorize and quantify a company's main accounts. The most used financial statements are the balance sheet, the income statement, the statement of origins and investment of funds, and changes in equity, in addition to the explanatory notes accompanying the aforementioned statements.

Financial Strategy – Management, carried out at the highest hierarchical level of a company, its internal funds, and the external funds to which it has access, with respect to investment processes, external growth, and significant changes in the shareholding of other companies.

Financial Transaction – Operation by which negotiations are carried out with financial securities or with credit transactions.

Financing Capacity – It is the denomination given to the balance of an agent's capital account, when it appears as an investment, that is, in assets.

Fiscal Anchor – Set of measures to keep the economy stable by containing public spending.

Fiscal Balance – Situation in which government expenditures are in line with revenues collected through taxes and other permanent sources of income.

Fiscal Deficit – Deficit resulting from the higher value of public expenditure in relation to the total collection of taxes by the government.

Fiscal Imbalance – Situation in which government spending exceeds revenue collected through taxes and other permanent sources of income.

Fiscal Policy – Government revenue and expenditure policy, which includes the tax burden on both individuals and companies, as well as the definition of government spending based on the amount of taxes collected.

Fiscal Stabilization Fund – Money retained by the Union that is no longer passed on to states and municipalities.

Fixed Cost – Cost that does not vary according to the volume of production, being, in general, contractual, as is the case, for example, of rent expenses.

Fixed Exchange – Exchange system in which a country's central bank establishes a fixed amount for the parity between the local currency and the dollar, or euro, for example.

Fixed Income – Denotes securities that pay, in defined periods, a certain remuneration, which can be determined at the time of investment or redemption.

Fixed Income Fund – Fund whose portfolio is basically composed of fixed income assets, or assets that behave as such, for example, derivatives.

Fixed Income – Income reported to the investor upon application, indicating how much he will earn and when he will have his money back.

Fixed Income Investment – Investment whose income value, usually expressed in the body of the security, is established in advance.

Fixed Rate – The one whose remuneration is determined at the time of application.

Flexibilization – Process by which rules, contracts, prices, among other elements, can be legally negotiated and modified.

Floating Exchange – This is the market through which remittances of profits from multinationals go abroad. The floating exchange rate also includes loan operations and the trade with jewelry and precious stones.

Flood the Market – Offering money to banks at low interest rates and, thus, stimulating a reduction in interest rates in general.

Foreign Capital – Capital available for the acquisition of productive or financial resources in one country by individuals or entities in other countries. External capital is also used.

Foreign Exchange Anchor – Set of foreign exchange policy measures that aim to keep the exchange rate fixed, indicating that the country has reserves to defend the currency against speculative movements.

Foreign Exchange Arbitration – Operation to buy a quantity of local currency and to sell another quantity of foreign currency, in order to obtain equivalence with the application of parity between them,

Foreign Exchange Band – Range or limit determined by the Government for the fluctuation of the real in relation to the dollar. The Brazilian system was adopted in March 1995.

Foreign Exchange Coupon – Defined as the interest rate in dollars in Brazil, calculated as the difference between the basic domestic interest rate (Selic) and the devaluation of the exchange rate, that is, the real against the dollar.

Foreign Exchange Crisis – A country's adverse economic situation, which can lead to a rapid loss of its foreign exchange reserves.

Foreign Exchange Market – Set of institutions that convert national currency into foreign currencies, and vice versa, in operations related to the external trade of goods and services.

Foreign Exchange Policy – Government economic policy that determines the value of the foreign exchange rate and the functioning of the foreign exchange market.

Foreign Exchange Regime – Exchange policy that provides for a band for the fluctuation of the real against the dollar.

Foreign Exchange Reserves – Assets in foreign currency and precious metals accumulated by a country, which act as a kind of insurance for the country to meet its obligations.

Foreign Exchange Risk – Possibility of risk to the investor caused by the instability of a country's foreign exchange policy.

Foreign Investment – Financial resources from abroad and applied to the economy of a country.

Foreign Trade – Trade in goods and services established between a country and its trading partners and which reflects the production and needs of each of them.

Formal Economy – Part of an economic system that respects the payment of taxes and the registration of employees and transactions.

Formal Employment – Employment that offers registration in the professional card for the worker.

Forward Market – Trading carried out on a stock or commodities exchange with a maturity agreed between the parties for a minimum of five days later. In general, maturities are 30, 60, 90, or 180 days.

Franchising – System that authorizes the sale of goods or services to a distributor or seller, who obtains the right to use a brand by paying a sum and a commitment to preserve the characteristics of that brand.

Free On Board – The abbreviation FOB is used in international maritime trade contracts, to ensure that the price of the traded goods covers all transport costs to the port of shipment, as well as all duties and taxes levied on the merchandise.

Free Trade Area of the Americas – Economic association between American countries, with the exception of Cuba, which aims to facilitate commercial transactions by reducing common customs tariffs. Known by the acronym FTAA.

Free Trade Area – Set of countries that organize the free movement of goods produced within their territories.

Free Trade – Free trade was initially described as an export and import activity by Josiah Child (1630–1699), one of the proponents of mercantilism, in 1668.

Free Zone – Area of a country where, by decision of the government, customs reductions are allowed, or tax benefits are granted for a certain period of time.

Funding – Determines the issuance of securities in the capital market, through which financial institutions, companies, and governments are able to obtain resources to finance their activities.

Fund Provider Shareholder – Small shareholder whose investment strategy is only the yield and appreciation of the shares.

Future Market – Market where you can buy and sell in the future. The shares can be purchased for settlement at a future date predetermined.

Futures Interest – Contracts traded on the Brazilian Commodities and Futures Exchange (BM & F), in which investors bet on the trend of rates in the future.

G-7 – The name was given to the periodic meeting, usually three times a year, of the seven richest countries: Germany, France, the United States, Japan and England, Italy, and Canada, with the aim of discussing relevant issues and decide, if applicable, coordinated actions.

GAP – Represents a gap in stock quotes. It is an interval in the quotation of the price of a given share, in which no business was registered in that price interval.

GATT – The General Agreement on Tariffs and Trade (GATT), along with the World Bank regulates the international financial system.

GDP – Gross Domestic Product is the sum of all goods and services produced in the country during the year.

GDP – Gross Domestic Product, or measure of the volume of goods and services produced in the economy in a given period. This index is formed by consumption, investment, government spending, stock level, and foreign trade balance.

General Liquidity – Financial analysis indicator, used to measure a company's liquidity, which shows how much a company has to receive in relation to what it owes, encompassing long–term assets and liabilities.

General Shareholders' Meeting – Maximum decision-making body of a corporation, which decides on mergers, takeovers, distribution of dividends to shareholders, after a proposal made by the board of directors

Gini Index – Index developed by sociologist and statistician Corrado Gini, as a means of measuring the division of wealth in societies. It measures the distribution of income (in some cases consumption expenditures) between individuals or households, within a deviation from a perfect distribution. Its value varies from 0 (or 0%), indicating perfect equality, to 1 (or 100%), which indicates perfect inequality.

Globalized Economy – Economic situation in which barriers to trade and the flow of capital between countries are eliminated.

Go Around – Informal or secondary auction of government bonds. It describes the process by which the Central Bank negotiating table asks its *dealers* to obtain buy and sell quotes.

Golden Share – Special title, owned by the government, that allows you to participate in decisions and administrative strategies of a company.

Gold Standard – The monetary system that linked the value of each country's currency to gold was suspended in 1931.

Goods and Services – Products of economic activity, or elements that constitute production.

Goodwill – Amount that the buyer overpays on the nominal value of a security.

Goodwill – Set of non-material elements arising from factors such as reputation, relationship with customers, and suppliers, or location, which contribute to the enhancement of a commercial establishment.

Grace Period – Defines the period in which the investor cannot redeem the funds invested in his investment plan or fund.

Gross Domestic Product – Sum of all goods and services produced in the country during the year. Known by the acronym (GDP).

Gross Financial Expense – Sum of interest expenses related to all financial obligations of a company, in the short or long term.

Gross Margin – Profit that the company makes before operating expenses.

Gross Revenue – That which occurs in the accounting period in which it takes place.

Guaranteed Capital – Investment fund modality, which aims to protect the initial investment in the event of a negative variation in the stock exchange index.

Guarantee Fund – Governmental fund that aims to guarantee the payment of loans granted by banking institutions to companies.

Hedge Fund – A fund that adopts strategies that cannot be used by traditional investment funds, but this does not necessarily imply that it is more or less risky.

Hedge – Operations designed to protect investors who want to reduce the risk of their investments, to avoid losses from asset price fluctuations.

Holding – Company that holds the controlling interest in another company, or in a group of subsidiary companies.

Hot Money – Short-term capital, which moves from one country to another, with possible currency exchange and speculative purposes.

Human Capital – Designates the body of knowledge and information accumulated by the organization's employees, as well as the investments destined to the educational training of these professionals.

Hyperinflation – Consistent increase in prices in the economy, or general lack of price control.

IBGE – The Brazilian Institute of Geography and Statistics is an important public company created to present studies and monitor the changes in macroeconomic accounts, such as unemployment, cost of living, inflation, average age, among others.

Ibovespa – Main index of the São Paulo stock exchange, which expresses the average daily variation of the trading on the São Paulo Stock Exchange. The Ibovespa was implemented in 1968 and is currently formed by a theoretical portfolio of 63 shares, which are chosen for their share in the market and their liquidity.

Incoterms – International rules for the interpretation of commercial terms.

Ideal State – The ideal state was conceptualized by Plato, around 380 BC, as the one in which the property belongs to everyone, and the work is qualified.

IDI – The Inclusive Development Index is a metric used by the World Economic Forum to assess countries' economic progress.

Idle Capacity – Difference between the actual volume of production and what would be possible to produce with the installed capacity.

IIF – Institute of International Finance. An organization that represents the largest 200 private banks in the world.

IMF – The International Monetary Fund started operating in 1945, headquartered in Washington, United States.

Import Rate – Percentage charged by the government on the price of a specific imported product.

Income Concentration – Process by which income, from profit, salary and other income converges to the same region, company, or restricted group of people.

Income Distribution – Process by which income, from profit, wages, and other income, is divided between regions, companies, or groups of people.

Income Received – Income from personal services, which generally includes employee salaries, tips, or other income.

Income Tax – Tax levied on income and wages, with different amounts for each type of income, and also in relation to individuals and companies.

Incorporation – Operation by which one or more companies are absorbed by others, which succeed them in all rights and obligations. An acquisition that implies the disappearance of the acquired company as a legally independent unit.

Increase in Productivity – Increase in the quantity of goods or services produced in the same area for a certain period.

Indebtedness Capacity – Ability that a company demonstrates to raise funds based on its financial structure. Typically, a company that has a ratio of equity to permanent assets less than 0.5 is at the limit of its debt capacity.

Indexing – It consists of linking the value of a capital or an income to the evolution of a reference variable, such as price, production, productivity, for example.

Indicative Targets – Flexible targets that, unlike quantitative targets, do not need to be achieved to release loans in the agreements signed with the IMF.

Inflationary Acceleration – Progressive increase in the rate of inflation.

Inflation – General phenomenon of adjustment, through monetary means, of the existing tensions in a socioeconomic group, which is characterized by the increase in the general price level and the depreciation of the currency.

Inflation Index – Index that reflects the percentage of the price increase during a given period.

Inflation Targets – Percentages that the government stipulates for the variation of the inflation rate.

Informal Economy – Part of an economic system, consisting of small groups of production, sales, or services, which does not respect the payment of taxes and the registration of employees and transactions.

Information Economics – Describes markets where a buyer has better information than another, according to George Akerlof.

Innovation – A change in products or production techniques.

INPC – The Brazilian Consumer Price Index is calculated by the Brazilian Institute of Geography and Statistics (IBGE) and considers the variation in prices in 11 regions: Rio, São Paulo, Belo Horizonte, Brasília, Porto Alegre, Curitiba , Belém, Fortaleza, Salvador, Recife and Goiânia. It is based on the budget of families with monthly income between one and eight minimum wages.

Insider – Investor who has access to information about a certain company, before they become known to the market.

Installed Capacity – It is the production potential of a given sector of the economy. To say that the industry is working with 70% of its capacity is the same as saying that it is with 30% of its idle production capacity.

Institutional Investor – Legal entity that manages third party funds, and whose investment composition is regulated by the State, with the objective of guaranteeing adequate and stable levels of profitability in the long term.

Insured Capital – Defines the amount of money fixed in the insurance policy

Intangible Assets – Characterizes the assets of a company, which do not have immediate physical representation, such as patents, franchises, names, and brands.

Intellectual Capital – Sum of the assets of a company, which represents the total of its knowledge and can guarantee a competitive advantage.

Intellectual Property – Any kind of property that comes from the conception or product of intelligence, to express a set of rights that belong to the intellectual, be it writer, artist or inventor, as the author of an imagined work , elaborated or invented.

Interbank Deposit Certificate – The IDC is a security that the bank launches to raise money in the market. Its function is to transfer money from one bank to another.

Interbank – Market in which transactions between banks are carried out. The Central Bank operates in this market to regulate the amount of money in the financial system. To do this, he takes or lends funds at a certain interest rate.

Interest on arrears – Also known as arrears interest, the term defines the interest rates charged by credit card companies in the event of late payment.

Interest Policy – Economic policy of the government that aims to keep the interest rate at a certain level.

Interest Rate – Cost of money in the market, regulated by the Central Bank. The interest rate is calculated by dividing the interest paid per year by the principal.

Interlocking Directorship – Cross participation in the board.

Interlocking Ownership – Cross ownership.

Intermediate Goods – Manufactured goods or processed raw materials used in the production of other goods.

Internal Debt – Debt assumed by the government with individuals and legal entities residing in the country itself.

International Economy – The international economy deals with the flow of goods, services, and capital through borders between countries.

International Monetary Fund – The IMF was created in 1944, by the Bretton Woods Agreement, and is the financial organism of the United Nations (UN) to correct imbalances in the balance of payments of member countries, to maintain the balance of the international economic system.

International Reserves – Assets in foreign currency and precious metals accumulated by a country. Similar to foreign exchange reserves

International Trade – Trade in goods and services established between countries and reflecting the production and the needs of each one of them.

International Travel – One of the items that make up the service account. On the expenditure side, tourist spending abroad on tickets, credit cards, accommodation, and dollars taken for travel are recorded. On the revenue side, the same expenses that foreign tourists incur in the country of origin are accounted for.

Inventory Turnover – Expresses how quickly the company is able to turn inventory over a year. The indicator is calculated as the quotient between the cost of goods sold and the value of the company's average inventory.

Investment Grade – Qualifies companies or governments with good payment capacity, those with investment grade.

Investment Account – Account type for people who have money invested, whose main characteristic is to exempt the investor from paying the tax in the movement of funds between accounts of the same title.

Investment Bank – Bank that offers loans and financing for setting up companies.

Investment Fund – Savings collection and investment organization, in which the variable capital is open to the public, and the value of the securities held by each participant is determined by the relationship between the total asset and the number of quotas, and not directly by the market.

Investment Fund – The most common form of financial investment, which functions as a condominium for individual resources of individuals or legal entities. In most cases, these funds operate as an open condominium, with no maximum number of participants, managed for the purpose of investing resources in the market and maximizing the return for the investor, called the shareholder.

Investments – New expenses that the Government intends to make in the country with a view to development, such as roads, railways, rural electrification, sanitation. They are accounted for in the General Budget together with financial investments, under the caption of other capital expenditures.

Invisible Hand – Adam Smith's most famous phrase, contained in the book An Inquiry Into the Nature and Causes of the Wealth of Nations, showed that he had a certain aversion to mercantilism and was, in fact, a criticism of the merchants' greed. In his own words, "Thus, the merchant or trader, driven only by his own interest, is led by an invisible hand to promote something that was never part of his interest: the well-being of society."

Invoice – Document that details all transactions, including debits and credits, that occurred in the reference period and that justify the amount to be paid by the credit cardholder.

IOF – The Brazilian tax on financial transactions is levied on the remuneration of all banking and financial activities, with the exception of interest itself.

IPCA – The broad consumer price index of Brazil, calculated by the IBGE, to measure the variation in inflation in households with incomes of up to 40 minimum monthly wages.

IPC-Fipe – Consumer Price Index calculated by the Economic Research Institute (Fipe) of University of São Paulo (USP), which considers the variation of prices in the capital of São Paulo.

IPC-RJ – Index that considers the variation in prices in the city of Rio de Janeiro, calculated monthly by the Getúlio Vargas Foundation (FGV), which is based on the expenses of families with income from one to 33 minimum wages.

IPO – Opening of the company's share capital to the stock market, or Initial Public Offer.

IIT – Individual Income Tax, which is due by people who had an income higher than the ceiling established by the Federal Revenue during the year.

Isoquant – Curve that represents various combinations of factors of production that result in the same amount of production. Analogous to the consumer indifference curve.

ISS – Tax on services of any nature, paid by all companies and self-employed workers who provide services.

Joint-Venture – Joint venture, which represents an association between companies or between countries, in the form of capital, labor, or natural resources.

Judicial Deposit – Sum of money deposited by a company, by order of the judiciary, in cases where the payment of a tax is questioned.

Know-how – Technological or human resources collection of a company, a country or a person.

Laffer Curve – The curve shows that the increase in taxes can cause a reduction in revenue. The idea was presented by Arthur Laffer, in 1974.

Landlord's Gain – Real gain obtained by the government in cases where the means of payment increase at a higher rate than physical production.

BGL – The Budget Guidelines Law establishes the rules for the preparation of the General Budget. It defines the percentage to be spent on investments, or how the resources will be distributed among the various states and municipalities. The BGL is approved by Congress in the first half of the year to guide the General Budget vote in the second half.

Lean the Market – Withdraw money from the sale of securities.

Leasing – Professional credit modality formed by a lease agreement for furniture or real estate equipment, accompanied by a promise to sell to the lessee.

Leverage – Effect of improvement caused by indebtedness on the profitability of a company's net worth. It represents the relationship between a company's indebtedness and its own capital, that is, its net worth.

Leverage – The use of borrowed money to supplement existing funds for investment purposes.

Liabilities – Comprises liabilities to be paid, that is, the amounts that the company owes to third parties, such as securities payable, accounts payable to suppliers, salaries payable, taxes payable, or mortgages payable. It represents the counterpart of the asset in the balance sheet of an economic subject.

Liberal Economy – In 1848, John Stuart Mill laid the foundations for the liberal economy, with the defense of trade and social justice.

Liberalism – Economic liberalism is the doctrine based on free competition and the individual initiative of economic agents.

Libor – in London Interbank Offered Rate, the interest rate charged on foreign currency loans, which is in effect in the London international financial market.

Limited Company – A company formed by at least seven partners, the capital of each corresponding to the proportion of the votes held and the responsibility of each partner, in case of bankruptcy or damage caused by the SA to third parties.

Limited Company – Company constituted by limited liability shares, which define the maximum amount of liability for each partner.

Liquidity – Ability to transform a particular asset or investment into cash. The volume of money circulating in the market. If liquidity is high, for example, there is a lot of money circulating through financial institutions. In the financial market, the term determines the capacity that a security has to be converted into currency.

Long Term Contract – Contract that establishes that a certain asset will be bought and sold in the future for a price fixed in the present.

Long-Term Debt – Indicator used in financial analysis, which serves to understand the capital structure of a company, calculated as the percentage of the company's invested capital composed of long-term funds from third parties.

Macroeconomics – A branch of economics that studies, on a global scale and by statistical and mathematical means, economic phenomena and their taxation in a structure or in a sector, verifying the relationships between elements such as national income, the level of prices, the interest rate, the level of savings and investments, the balance of payments, the exchange rate and the level of unemployment. The first analysis of an entire economy was carried out by François Quesnay, in 1758.

Majority Shareholder – One who owns, at least, more than half of the voting shares of a company, and controls it.

Marginal Analysis – Comparison between the costs incurred and the benefits obtained from some financial strategies, so that the company can better analyze its strategy in an attempt to maximize its profitability.

Marginal Cost – Concept used in economics to describe the changes caused in the total cost for a unit change in the quantity produced.

Marginal Utility – The satisfaction that an individual experiences from the consumption of an additional unit of a certain good or service.

Marketing – Market research to plan for possible future product launches, taking into account existing or possible needs and the company's research and adaptation prospects.

Market Opening – Economic policy that provides for the entry of products, companies and foreign investments in a country.

Market Reserve – Situation in which a productive sector is protected by the government through restrictions on imports and stimulating production in the national territory.

Market Share – Market share of a company, or group, within its segment.

Market Value – Value referring to the sum of all the shares of a company, which is different from the book value, as it does not take into account factors such as the net debt and the assets of the corporation.

Mathematical Economics – Application of mathematical methods to represent economic theories and analyze problems proposed by Economics, with the aim of formulating or deriving theoretical relations in a generic, rigorous and simple way. Augustin Cournot (1801–1877), a French mathematician and economist started this area of research with the publication of the book *Researches sur les Principles Mathématiques de la Theorie des Richesses*, in 1838.

Means of Payment – Resources considered immediately available to the population. It is measured by the money held by the public, plus demand deposits at commercial banks.

Medium of Exchange – The drop item accepted in transactions for goods and services. It is also one of the functions of money.

Merger – The business union of two or more companies. Eventually, the shares of the companies are exchanged for shares of a third company, which results from the merger.

Microeconomics – Theoretical process designed to determine the general equilibrium conditions of the economy based on the behavior of individual economic agents, producers, and consumers. It deals with the way in which the individual entities that make up the economy, private consumers, commercial companies, workers, large landowners, producers of private goods or services act reciprocally.

Minimum Wage – Lowest wage fixed by law, in order to guarantee wage earners of the least favored categories an income corresponding to the vital minimum, defined in relation to a given social environment.

Minority Shareholder – One who owns a number of shares with voting rights less than half, which may or may not hold control of the company, depending on the voting power of the other shareholders.

Model – Formal representation, usually in a system of equations, a logical system or an algorithm, which forms a coherent set of relationships between economic phenomena.

Monetarism – Theoretical chain that attributes money to a determinant role in economic fluctuations, whose supporters defend the quantitative theory of money and its implications.

Monetary aggregate – Set of homogeneous elements that make up the money supply in a country.

Monetary Anchor – Set of measures that aim to keep the economy stable by determining a limit on the money supply.

Monetary Authority – Federal institution responsible for establishing the rules that govern a country's monetary and financial system.

Monetary Correction – Periodic readjustment of certain prices in the economy by the value of past inflation, in order to compensate for the loss of the purchasing power of the currency.

Monetary Offer – Set of credits formed by monetary and quasi-monetary availabilities.

Monetary Policy – Control of the amount of money in circulation in the market, which allows the definition of interest rates.

Monetary Stability – Situation of equilibrium in the value of currency achieved in a country through the control of the money supply, interest rates and the public deficit.

Monetary Standard – Value, or matter, conventionally adopted as the basis of the monetary system of one or several countries, and in relation to which the other types of currency and the monetary units will be defined.

Monetary Zone – Set up, following a formal agreement or as a result of a de facto state, by a group of countries or territories, which observe particular rules in their monetary relations and confer on the currency of the principal of these countries an essential role in internal payments in the area and with the rest of the world.

Monetization – Increased use of paper money or checks.

Money Market – Set of institutions that carry out short and very short-term credit operations to cover the momentary disengagement of economic agents, including the National Treasury and the banks themselves. It serves as the main instrument for regulating the economy's liquidity for the Central Bank, through the purchase and sale of government bonds.

Money – The stock of assets used for transactions.

Monopolistic Competition – When two or more companies that produce similar goods or services, but are not substitutive for each other, maintain a certain degree of control over the prices of these products.

Monopoly – Situation of a market in which there is no competition in the offer.

Moratorium – Provision that suspends payment within a period fixed by law or under a contract.

Mortgage Bill – Bond issued by financial institutions authorized to grant mortgage credits. The profitability of this type of investment is linked to the nominal value of real estate financing, adjusted for inflation or variation of the Interbank Deposit Certificate (CDI), and can be pre-fixed, floating and post-fixed.

Mutual Fund – Gathering resources from several investors, called quota holders, managed by a broker or investment bank and invested in shares, bonds, and securities. The yields are distributed according to the proportion of the shares.

National Income – Aggregate representing the flow of national resources in goods and services, generated over a given period. It includes salaries, income from self-employed professionals, private profits and profits obtained by public companies, interest, rents, and income from leasing.

National Monetary Council – Main body of the Brazilian Financial System, the NMC was created by Law 4,595 of 1964. Its competence as a disciplinary body of the Capital Market was established by Law 4,728 of July 14, 1965, having the purpose of formulating the currency and credit policy.

National Product – An aggregate that brings together all the products from the different branches of a national economy during a certain period, usually one year.

National Treasury Notes – Government bonds, issued by the National Treasury.

Neo-liberalism – Political-economic doctrine that constitutes an adaptation of the principles of liberalism, in which individual activities and initiatives are encouraged as opposed to those of the State.

Net Debt – Sum of all financial obligations, as loans, debentures, fixed income securities, of a company, short or long term, less the company's cash, that is, the sum of the instruments that can be considered as paper money.

Net Financial Expenses – Sum of interest expenses related to all financial obligations of a company, discounting any interest income that the company may have from its financial investments from this amount.

Net Funding – Difference between deposits and withdrawals related to an application for a specified period.

Net Gain – In the cash market, it is equivalent to the difference between the sale value of a given financial asset and its acquisition cost. In the futures market, it is the result of the sum of the daily adjustments that occurred in each month.

Net Income – Income of an individual or entity, over a specified period, less taxes and social contributions.

Net Investment – The amount of investment, after depreciation of the depreciated capital. It also indicates the change in the capital stock.

Net Margin – Indicator used in the financial analysis of companies, which expresses the relationship between the company's net profit and its net sales revenue.

Net Profitability – Percentage of income from financial investments, less taxes and fees.

Net Revenue – Sales revenue, less sales (NR) and production (PT) taxes, returns and discounts.

Net savings – Gross savings less amortization of fixed assets.

Net – What is readily convertible into a medium of exchange, or easily used to carry out transactions.

Net Working Capital – Difference between current assets and current liabilities, which are included in a company's balance sheet on a given date.

Niche Market – Group of consumers who share the same interests and objectives and constitute a high potential for consumption.

Nominal Deficit – It is the concept of public deficit that, in addition to revenues and expenses includes the expenses with the payment of interest on the public debt.

Nominal Income – Gain obtained on financial investment, without discounting any inflation rates.

Nominal Interest – The interest rate that includes the monetary restatement of the amount borrowed.

Nominal Shareholder – A company that appears as the owner of shares, but which only covers up the effective shareholder who does not wish to identify himself.

Nominal Value – This is the share price, mentioned in a company's registration letter.

Nominal – What is measured in current monetary value, not adjusted for inflation.

Nominative Share – The one whose certificate is nominal to its owner. The certificate, however, does not characterize the possession, which is only defined after the issuance is registered in the Book of Registered Shares of the issuing company.

Non-Financial Assets – These are fixed assets, which participate in several production cycles, and those in circulation, which are consumed or transformed in a specific production or distribution cycle.

Non-taxable income – Gain exempt from income tax collection.

North American Free Trade Agreement – Economic association established between the United States, Canada and Mexico, to facilitate and increase commercial transactions between these countries through the progressive elimination of customs tariffs. Its acronym NAFTA.

NTN – The National Treasury Note is a paper that can be used to cover holes in the federal budget or to exchange foreign debt, in dollars, for domestic debt.

OECD – The Organization for Economic Cooperation and Development was established in 1961 and is concerned with a wide variety of economic matters (Kreinin, 1987).

Offer – Making goods or services available to the market, usually for a fixed price and time.

Offshore – Denomination given to purchases made by the American government abroad, within the scope of its international aid policy.

NGB – The Nation's General Budget details all expenditures that the government is authorized to make during the year.

Oligopoly – Market structure in which a small group of companies controls a significant portion of the offer of products and services.

Oligopsony – Situation of a market in which competition is imperfect on the demand side, due to the presence of a very limited number of buyers.

Open Capital – A company whose securities, such as shares, for example, are registered with the Securities and Exchange Commission (SEC), and which are admitted to trading on the market for bonds and securities, on the stock exchange or over the counter.

Open Economy – One in which there is freedom for imports and mobility of factors across geographical borders.

Open Market – Market in which monetary authorities, such as the central bank, operate with government bonds, in order to regulate and control the means of payment, while financing the internal federal debt.

Operating Cost – Sum of expenses necessary to maintain productive services of a private or state owned company.

Operating Expense – Sum of all costs and expenses incurred by the company in the course of its activities.

Operating Margin – Quotient between the operating result and the company's net sales revenue.

Operating Profit – Profit from the companies' productive operations.

Operational Deficit – Deficit resulting from the higher value of current expenses, investments, and interest in relation to current revenues.

Optimize – Achieve the best possible result, subject to a set of restrictions.

Option – Contract traded in the financial market that entitles, upon the immediate payment of a premium, to buy or sell financial assets in installments.

Options Market – The stock option is a type of contract that guarantees its holder the right to buy a lot of shares at a price fixed in that contract, for a specified period. The contract counterpart, the writer, undertakes to sell the lot at the fixed price, until the expiration date, if the holder wishes to exercise his right.

Ordinary Share – Title that grants its holder the right to vote at a shareholders' meeting.

Organicism – Doctrine that compares the economic system to a living being.

Outsourcing – Process of transferring services to third parties, usually consisting of companies or specialized individuals.

Overnight – Financial investments made in the open market in one day to be redeemed the next day.

Paid-in Capital – Denotes a company's capital account, in which the value of the subscribed shares has been fully received.

Parallel Exchange – It is the market that exists when the country does not have a totally free exchange rate.

Parent Company – Company that controls another company, being also called in holding company.

Pareto, Great from – Formulated by Vilfrido Pareto, in 1906, it is the state in which no individual can be in a better situation, without another being in a bad situation.

PAR – Share value identical to the official or nominal share.

Patent – Document issued by the government and granted to an inventor or its representatives, in order to protect their property and exploitation rights from an industrial invention.

Paycheck – A denomination used for the salary receipt.

Peer-to-Peer – Computer network configured to allow sharing of files and folders between all, or only a few, selected users. Peer-to-peer networks are common in small offices that do not use a dedicated file server.

Pension Fund – Supplementary pension plan that is sponsored by unions, class associations, professional councils, and cooperatives, and that guarantees the payment of social security benefits complementary to those paid by social security.

Per Capita Income – Result of dividing the total amount of taxable income by the number of people, in economics, an indicator used to measure the degree of development of a country.

Performance Criterion – Set of goals of the agreement with the International Monetary Fund (IMF), by which a country is evaluated. Failure to meet these targets means the interruption of transfers of loan installments and demands a new renegotiation.

Permanent assets – Sum of a company's fixed assets, such as real estate and machinery, with long-term investments, such as participation in associated companies.

PPR – The Price/Profit Ratio is the number of years that it would take to recover the capital invested in the purchase of a share, through the receipt of the profit generated by the company.

Phillips, Curve – A relationship between inflation, cyclical unemployment, expected inflation and supply shocks, derived from the short-term aggregate supply curve.

NDP – Net Domestic Product, the measure that relates the amount of wealth of a given nation, state, or municipality, with its productive capacity.

Pledge – A movable property belonging to a debtor and delivered to his creditor to ensure the settlement of his life.

Policy – Denotes the most important document when insurance is taken out.

Portfolio – Set of securities, shares, and other assets, launched on the international or domestic market, which makes up the assets of an economic agent.

Post-fixed Income – Income that pays for monetary restatement in the investment period, plus interest, on the adjusted value of the investment. In this application, the investor only knows what his return will be on the security's maturity.

Post-fixed security – The one whose yield is determined by the variation of a certain index plus an interest rate determined at the beginning.

Preferred Share – An action that gives its holder priority in the receipt of dividends, or in case of dissolution of the company, in the repayment of capital. They do not give the right to vote at the company's meetings.

Premium – Previously agreed indemnity, which the forward buyer of a stock exchange price pays to the seller on the settlement day, in the event of the withdrawal of an already contracted operation.

Price – Monetary value expressed numerically associated with a commodity, service, or asset. Regardless of its objective use value and subjective satisfaction value, the price of a good or service only exists to the extent that it is within an exchange relationship.

Primary Auction – Sale of public securities, with the publication of a public notice, for the entire market, which is also known as a formal auction.

Primary Deficit – Negative result of public accounts, which includes the National Treasury, Social Security, and Central Bank, disregarding interest on public debt.

Prime Rate – The most important interest rate in the world economy, charged by American banks to their main customers.

Private Company – Company, belonging to individuals or groups, which aims to produce goods and services for profit.

Private Initiative – Set of activities carried out by economic agents outside the governmental scope.

Private Property – Aristotle argued, around 350 BC, in favor of private property, but against the accumulation of money as the sole objective.

Private Sector – Sector of the economy, represented by private companies, whose decisions are determined by the market.

Privatization – Process by which the controlling interest in a company or financial institution belonging to the government is transferred to the private sector, whether for individuals or companies.

Procyclical – which moves in the same direction as production, income and employment during the business cycle.

Production Capacity – Volume of goods or services that a company can produce during a predetermined period of work.

Production Cost – Sum of expenses related to raw materials, employed labor and other expenses resulting from the elaboration of a product.

Production – Creation of a good or an adequate service to satisfy a need.

Production Factors – These are the services associated with resources, or productive agents, not the resources themselves. Although there are many factors of production, it is convenient to group them into some categories, such as labor, capital, and property services.

Production Goods – In addition to capital goods, they also refer to intermediary and raw materials.

Productive Sector – Economy sector that is responsible for the production of goods and services in general.

Productivity – Determines the efficiency of a company or organization in the use of resources. The productivity of a company can be calculated by dividing physical production, obtained in a unit of time, by one of the factors of production: labor, goods, capital.

Product Life Cycle – Reflects the steps prior to the product's arrival at the production line, including design, development, prototyping, and testing, followed by customer use of the product, disposal, or recycling.

Profitability – Expresses the appreciation, or devaluation, of a given investment in percentage terms.

Profitable Assets – For financial institutions it reflects the sum of all assets that generate a financial return for the institution. The total return on these assets is included in the institution's gross income from financial intermediation.

Profit – Gain received from an operation in a given period of economic activity. It can be gross when considering variable expenses, such as taxes, or net, when taxes are discounted.

Profit Margin – Percentage of the gain obtained by a company in relation to its total sales.

Profit Sharing – Fraction of the profits of a company, to be distributed, in addition to the part from the first dividend and, possibly, interest, destined to the board of directors or the fiscal council as a complementary remuneration.

Pro-labore – Remuneration for a company to its partners for the work performed in the administration.

Promissory Note – Document issued by the debtor, who is obliged to pay his creditor, or his order, a certain amount, on a defined maturity date.

Property, Plant, and Equipment – Comprised of the sum of tangible assets used in the company's operational activities and which must not be converted into cash, or consumed in the course of its activities, by example real estate, machinery, equipment, land.

Pro Rata – Latin expression that has a sense of division, or proportional measurement.

Protectionism – Doctrine, theory, or economic policy that advocates, or puts into practice, a set of measures that favor domestic activities and penalize foreign competition.

Public Assets – Resources, in various forms, belonging to the government.

Public Bond – Bond issued by the government, which bears interest to the bearer at the time of redemption.

Public Capital – Resources owned by the public sector or state owned companies.

Public Debt – Everything the government spends on loans and securities issues.

Public Debt Rollover – Refinancing of issued papers to cover gaps in the government budget.

Public Deficit – Negative result of public accounts, resulting from the excess of public spending in relation to the funds raised by the government.

Public Finance – Economy sub-area that studies aspects related to the revenue and expenditure flows of public sector activities.

Public Fund – Set of monetary resources, owned or held, public or private, whose movement is not determined by an exclusively private logic, but includes some compulsory element, incentive or regulation established in the public sphere.

Public Non-Financial Sector – Comprised of federal, state, and municipal public companies, except banks, securities distributors, brokers, and other companies with permission to operate in the financial market.

Public Sector Primary Surplus – When the government achieves that its total revenue exceeds its expenses, less interest expenses and monetary correction of debts.

Public Tender Offer – Intention made public, by a potential control group, for the purchase of shares in a company, generally, defining conditions, such as price, minimum or maximum quantity to be acquired. The objective is to significantly increase the shareholding with voting rights, typically aiming at control.

Purchasing Power Parity – The doctrine, according to which, goods must be sold at the same price in any country. This implies that the nominal exchange rate reflects differences in price levels. This parity adjusts the GDP and per capita income of the poorest countries, considering that services and goods in these countries have lower costs.

Quality Control – Administrative strategy developed with the participation of the company's human resources, which aims at customer satisfaction, by offering superior quality products or services.

Quantitative Targets – Targets agreed with the directors of the IMF and expressed in reais and not in percentages, as is the case of the public sector primary surplus.

Quantitative Theory of Currency – The doctrine that emphasizes that changes in the amount of money lead to changes in nominal expenditures.

Quota – Or quota, is the fraction of an investment fund. Therefore, the equity of a fund is the sum of shares that were purchased by different investors.

Raider – Individual or legal entity that becomes a company acquirer without the agreement of its board of directors, using the public offer technique.

Rating – Note that international credit rating agencies assign an issuer, which can be a country, a company, or a bank, according to its ability to pay a debt.

Real Interest – Rate on a loan, or financing, not including monetary restatement of the amount borrowed.

Real Plan – Brazilian stabilization plan implemented in, July 1994, by president Itamar Franco, based on the creation of a stable currency, the real.

Receivables – Denotes all assets that a company is entitled to receive, such as promissory notes.

Recession – Decrease in economic activity, with a fall in production and an increase in unemployment, which occurs when the volume of wealth that a country produces (GDP) decreases in relation to what it produced in the previous year.

Rediscount – Monetary policy instrument used by the central bank to regulate the banking system's liquidity system used when commercial banks, despite all their cash forecasts, need cash reinforcement or are uncovered in clearing check. In these cases, the bank issues a promissory note in favor of the Central Bank and receives credit in its deposit account at Banco do Brazil.

Remuneration – What is perceived by an individual, or by a community, as a result of capital or as a result of work.

Rentier – Individual who earns income from capital investment.

Repurchase – Commitment made by a financial institution to re-purchase the security traded, at a future date, before the security matures.

Retail Bank – Commercial financial entity, which carries out transactions directly with individual consumers and not with companies or other banks.

Revenue – Sum of all amounts received in a given period of time. For a company, revenue is formed by sales, the part received for credit and any income from financial investments.

Risk Analysis – Continuous and systematic assessment of adverse effects, or risks, that may affect a particular company in a competitive market.

Risk Aversion – The propensity of the investor to pay an intermediary, for example, an insurance broker, to avoid an uncertain prospect.

Risk – Element of uncertainty that can affect the activity of an agent or the course of an economic operation, to which a probability distribution can be attributed.

Salary Isonomy – Principle that guarantees equal salary adjustment for civil servants who perform the same duties in different segments of the public sector.

Salary – The amount paid for a unit of work.

Savings Account – Account in a bank that offers monthly remuneration guaranteed by the government, in interest and monetary restatement, for the amounts deposited.

Savings – Destination given to monetary income not used for consumption, either through hoarding, investment, loan or for future direct investment.

Seasonality – Denomination of the period of the year with the highest activity in a given sector of the economy. The industry has a higher level of activity in the months of September and October when production increases to meet the orders of the trade for Christmas sales.

Secondary Auction – Sale of securities only to some financial institutions, which they can then pass on to others. Also called an informal auction.

Secondary Cash Surplus – When the investor gets his net income to exceed his expenses. Savings, measured as a percentage in companies as a contribution margin, are directed towards investments with rates of return compatible with future cash needs.

Second Line – Shares of smaller companies, both due to their size and the volume of shares traded, but with good economic and financial performance.

Securities and Exchange Commission – The SEC supervises and regulates the market for variable income securities, in addition to ensuring efficient and regular operation of the stock and over-the-counter markets, and protecting securities and market investors.

Securities Debt – Volume of securities that the Government issued and sold to the market.

Securities – Securities representing securities, such as common or preferred shares and government bonds.

Securitization – Process by which assets are transformed into securities.

Securitized Debt – Title of responsibility of the National Treasury issued as a result of the assumption and renegotiation of the Union's debts or assumed by it under the law.

Security Deposit – Commitment made by someone to take responsibility for fulfilling an obligation underwritten by another person, in case it fails. The beneficiary of the security is called the principal debtor.

Self-financing – Portion of investments that are funded from the accumulated profits of a company or economic group.

Selic – Abbreviation of the Brazilian Central Bank's Special Settlement and Custody System, where all debit and credit transactions made between banks and other accredited financial institutions are recorded.

Sensitivity Analysis – Analysis of the effects that the change of some variables can have on the projection of the results of a company, in order to measure the degree of variation of these results in relation to the changes of these variables.

Service Account – Records income and expenses with international travel, expenses on debt, contracting freight and insurance, and remittances of profits and dividends from Brazilian companies.

Services – Provision of assistance or carrying out tasks that contribute to the satisfaction of individual or collective needs, in any other way than by transferring ownership of a material good.

Share – Document of ownership of a fraction of the capital of a company. The capital of a company can be divided into equal fractions, allowing each partner to subscribe for shares according to their availability of capital.

Shareholders' Equity – Equity value that reflects the sum of paid-in capital, capital reserves, revaluation reserves, profit reserves, and accumulated profit or loss for the period. The total assets of a company are equivalent to the sum of all its liabilities plus its shareholders' equity.

Shareholding Control – Defines the decision-making power of one or more shareholders over a given company and this control is guaranteed by the possession of the largest number of voting shares.

Single Rate – Single percentage charged on income tax for all taxpayers.

Social Capital – Share of financial resources placed on the company by its shareholders.

Socialism – Doctrine that advocates the organization of an egalitarian society, free from exploitative relations between social classes, and that ensures the primacy of collective interest over individuals.

Social Security Deficit – It is the difference between what the government collects with the contribution of civil servants and what it pays as benefits to active and inactive public servants.

Society – Legal entity, instituted by a contract, which brings together several people who are obliged to jointly use values, goods, or work, for profit.

Speculative Attack – Situation in which a country suffers the action of investors, who apply in the local currency to then exchange it for dollars, which causes the currency to devalue or raise the interest rate.

Speed of Money – The estimated number of times per year that the average value of a currency of the money stock is spent.

Spending – Transformation of savings into consumption.

Spin-off – Process of transfer, by a company, of portions of its assets to one or more companies, existing or constituted for this purpose, extinguishing the spun-off company, if there is a version of all its assets.

Split – Split of shares. A split represents the distribution by the company of a certain number of papers for each existing one.

Spot Market – Market in which physical settlement, delivery of securities by the seller, takes place on the second business day after the deal is closed, and financial settlement, payment of securities by the buyer, takes place on the third business day after the negotiation, only through the actual physical settlement.

Spread – Bank margin added to the rate applicable to a credit. The spread varies according to the borrower's liquidity and guarantees, the volume of the loan and the redemption term.

Stagflation – Economic situation characterized by the conjunction of a tenet to stagnation, or recession, accompanied by inflation.

Stagnation – Factor resulting from the demand for investment, exports, consumption, or economic activity in general, with an impact on production.

Stand By – Agreement between the International Monetary Fund (IMF) and a member country, authorizing it to make, for a certain period and for a certain amount, withdrawn from the fund, in foreign currency.

Stock Exchange Index – The SEI that measures the daily variation in the prices of the most traded shares on the Stock Exchange.

Stock Exchange – Non-profit civil association, with the objective of maintaining a local or electronic trading system suitable for carrying out securities purchase and sale transactions. The Stock Exchange also aims to preserve high ethical standards of negotiation and to disclose the operations executed with speed, breadth, and details.

Stock Market – Segment of the capital market that comprises the primary placement of new shares issued by companies and the secondary interest, on stock exchanges and over the counter, of shares already put into circulation.

Structural Unemployment – Unemployment resulting from wage inflexibility and job rationing.

Subcontracting – Any contract that implies the provision of goods and services by one company in relation to another.

Subscribed Capital – Amount effectively subscribed or deposited, in the company by its shareholders.

Subscription – Launches of shares, debentures, or other securities by financial institutions that undertake to advance capital, or to acquire part of the securities not absorbed by the market or, at least, to make efforts to minimize any problems with the sale of securities.

Subsidiary – Company controlled mainly by another company.

Subsidiary – Company whose majority voting rights are concentrated in another company, referred to as a parent. Also called a subsidiary.

Swap – Exchange between different currencies and carried out between banks through a cross game of deeds, with prior agreement and redemption clause. It is drawn on credit, and the right to withdraw is then reconstituted in a short period of time

Tariff – Rate applied to imported goods

Tax Adjustment – Set of procedures of the Federal Government in order to spend less than collected.

Taxation – System by which the Federal Government, the States or the Municipality levies and collects taxes.

Tax Avoidance – Operation by which taxpayers withhold information in order to pay less taxes.

Tax Charge – Amount of taxes collected in the country. The load is measured as a percentage of the GDP.

Tax Evasion – Process by which the payment of taxes is withheld. It is equivalent to tax evasion.

Tax Exemption – Exemption from the payment of taxes legally granted to some products, services, or activities.

Tax Exemption – Procedure that allows the reduction of the amount of the tax levied on an economic transaction, or due by an economic agent.

Tax Gain – Advantage obtained by the government as a result of increased taxes or reduced expenses.

Tax Haven – Economic zone where the fiscal and monetary regulation of banking activities are light, or nonexistent.

Tax Incentive – Incentive by which the government offers tax exemption or reduction for companies that invest in activities that are important for a country's economic policy.

Tax Revenue – Revenue collected by the government through taxes, fees and contributions.

Tax Waiver – Tax exemption granted by the government with the objective of developing a region or stimulating certain sectors of the economy.

The Capital – The first volume of *Das Kapital*, by Karl Heinrich Marx, was published in 1867.

Term – Synonymous with a term for financial investments.

Term Deposit – Bank deposit, with restricted movement, which yields interest to the account holder.

Third-Party Capital – Portion of the total capital invested in the company that does not belong to the shareholders, being, in general, equivalent to the company's debt.

Title – Document of ownership of a good or value that is not in the possession of the holder. They are divided into short-term commercial securities. Your buyer is entitled to receive the amount represented by him, for example, bill of exchange, promissory note, or duplicate.

Tourism Account – Shows the expenses of Brazilians on international travel and the income obtained by the country when foreigners visit Brazil.

Trade Balance – Balance resulting from the difference between imports and exports.

Trade Deficit – Reflects the difference between what the country collected from exports and what it spent on imports. When the result is negative, and imports are greater than exports, it characterizes a trade deficit. If the result is positive, a trade surplus occurs.

Trade Opening – Government policy of phasing out tariffs and restrictions non-tariff duties on imported products.

Transaction – Any act by which an economic unit manifests its participation in economic life.

Treasury Bill – A government-issued, fixed-term security that pays a market interest rate, are also known as government bond.

Trial Balance – Partial balance of a company's economic situation and equity, which refers to a specific period of the company's fiscal year.

Tribute – Revenue instituted by the Union, the States, the Federal District and Municipalities, which includes taxes, fees, and improvement contributions, under the terms of the Constitution and the laws in force in financial matters.

Trust – Assets held by individuals or companies for the benefit of third parties.

Tulip – The speculative bubble in the Dutch tulip market burst in 1637, leaving thousands of investors ruined.

Uncertainty – Condition in which there is no probability distribution that can be associated, due to the mutability of the cause of the data changes over time.

Underground Economy – Economic transactions that are hidden, with a view to avoiding tax evasion or to hide illegal activity.

Underwriters – Financial institutions specializing in the launching of shares and debentures in the primary market.

Unemployment Rate – The percentage of those in the workforce who have no jobs.

Usury – Practice consisting of charging interest rates higher than usual or those permitted by law, when borrowing.

Utilitarianism – Theory formulated by Jeremy Bentham, in 1791, whose aim is to provide the greatest happiness for the greatest number of people.

Utility – Quality of what is used by the economic agent. The utility is one of the basic notions of the economy, such as that of value.

Value-Added – The value of a company's product, less the value of the intermediate goods that the company acquired.

Variable Cost – Cost that varies directly with the company's sales volume. Production costs are examples of this type of cost.

Variable Income Fund – In addition to equity funds, this category also includes foreign exchange funds, derivative funds, and external debt funds.

Variable Income – Security whose remuneration is not determined in advance. The profitability of these applications depends on market conditions.

Variable – Size that can vary in its own way or as a function of other variables. In descriptive statistics, quantitative, discrete, or continuous feature.

Venal Value – Estimate performed to define the value of an asset according to the government's view.

Venture Capital – Portion of the company's capital that is invested in activities or instruments in which there is a possibility of losses or gains higher than those normally expected in the company's usual activities.

Venture Capital – Source of financing for new businesses, or for business recovery, which generally combines high risk with the potential for high return.

Volatility – Sensitivity of the price of a stock or a portfolio to changes in stock exchange prices.

Warrant – Negotiable security issued by warehouse companies that represent the goods deposited there. In the capital market, it is a document that guarantees to shareholders, with a fixed term and price, the acquisition of additional lots of shares.

Wealth – The term refers to some accumulation of resources. John Locke concluded that wealth comes from work, not trade, in 1689.

Workforce – Those workers in the population who are employed, or are looking for work.

Work – Human talent, whether physical or mental, which is applied to the production of goods and services.

Working Capital – Defined as the difference between a company's current assets and current liabilities. The portion of an entity's investment that is always being used and renewed to allow for ongoing operations.

Working Capital – Part of the capital that is used to finance the company's current assets, and that guarantees a safety margin in the financing of the operating activity. It refers to the capital, owned or of third parties, used by the company to finance its production, for example, the money used to pay suppliers. Third-party funds are, in general, raised with commercial banks through operations such as a duplicate discount.

World Bank – International financial organization in developing countries. In conjunction with the General Agreement on Tariffs and Trade (GATT), the World Bank regulates the international financial system. It was created in 1944 and gave rise to the World Trade Organization.

Write-off – Transfer the total balance of an asset account to a loss or expense account.

Yield – Financial return on a good or service performed. The actual rate of return for the investor, or effective cost for the issuer, of a guarantee for a specified period of time.

Bibliography

Alencar, M. S., **A Quem Serviu a Privatização das Telecomunicações?**. Artigo para jornal eletrônico na Internet, Jornal do Commercio *On Line*, Recife, Brasil (2007).

Alencar, M. S., **Diferenciais econômicos para o desenvolvimento do País**. Artigo para jornal eletrônico na Internet, Jornal do Commercio *On Line*, Recife, Brasil (2008a).

Alencar, M. S., **Economia móvel**. Artigo para jornal eletrônico na Internet, Jornal do Commercio *On Line*, Recife, Brasil (2008b).

Alencar, M. S., **Relatividade para invisuais - II**. Artigo para jornal eletrônico na Internet, Jornal do Commercio *On Line*, Recife, Brasil (2008c).

Alencar, M. S., *Probabilidade e Processos Estocásticos*. Editora Érica Ltda., ISBN 978-85-365-0216-8, São Paulo, Brasil (2009).

Alencar, M. S., **A volta Triunfal do Disco Analógico – 1**. Artigo para jornal eletrônico na Internet, Jornal do Commercio *On Line*, Recife, Brasil (2010).

Alencar, M. S., **As Empresas que Dominam o Setor de Telecomunicações no Brasil**. Artigo para jornal eletrônico na Internet, NE10 – Sistema Jornal do Commercio de Comunicação, Recife, Brasil (2011a).

Alencar, M. S., **Presidentes não Recebem Propinas, Ministram Palestras – 1**. Artigo para jornal eletrônico na Internet, Jornal do Commercio *On Line*, Recife, Brasil (2011b).

Alencar, M. S., **Presidentes não Recebem Propinas, Ministram Palestras – 2**. Artigo para jornal eletrônico na Internet, Jornal do Commercio *On Line*, Recife, Brasil (2011c).

Alencar, M. S., **Presidentes não Recebem Propinas, Ministram Palestras – 3**. Artigo para jornal eletrônico na Internet, Jornal do Commercio *On Line*, Recife, Brasil (2011d).

Alencar, M. S., **Presidentes não Recebem Propinas, Ministram Palestras – 4**. Artigo para jornal eletrônico na Internet, Jornal do Commercio *On Line*, Recife, Brasil (2011e).

Alencar, M. S., **O maior Item da Pauta de Exportação do Brasil dá Prejuízo – Parte I**. Artigo para jornal eletrônico na Internet, NE10 – Sistema Jornal do Commercio de Comunicação, Recife, Brasil (2012a).

Alencar, M. S., **O Mundo da Economia**. Artigo para jornal eletrônico na Internet, NE10 – Sistema Jornal do Commercio de Comunicação, Recife, Brasil (2012b).

Alencar, M. S., **O que o País Perdeu com a Venda da Embratel**. Artigo para jornal eletrônico na Internet, NE10 – Sistema Jornal do Commercio de Comunicação, Recife, Brasil (2012c).

Alencar, M. S., **O Universo Econômico**. Artigo para jornal eletrônico na Internet, NE10 – Sistema Jornal do Commercio de Comunicação, Recife, Brasil (2012d).

Alencar, M. S., **Privatização do Bem**. Artigo para jornal eletrônico na Internet, NE10 – Sistema Jornal do Commercio de Comunicação, Recife, Brasil (2012e).

Alencar, M. S., **A Densidade da Riqueza**. Artigo para jornal eletrônico na Internet, NE10 – Sistema Jornal do Commercio de Comunicação, Recife, Brasil (2013a).

Alencar, M. S., **Capitalismo para Leigos**. Artigo para jornal eletrônico na Internet, NE10 – Sistema Jornal do Commercio de Comunicação, Recife, Brasil (2013b).

Alencar, M. S., **Como a Falta de Investimento em Comunicações parou o Brasil**. Artigo para jornal eletrônico na Internet, NE10 – Sistema Jornal do Commercio de Comunicação, Recife, Brasil (2013c).

Alencar, M. S., **Sobre Colonialismo e Privatização**. Artigo para jornal eletrônico na Internet, NE10 – Sistema Jornal do Commercio de Comunicação, Recife, Brasil (2013d).

Alencar, M. S., **Sobre Privatizações, Tarifas e Impostos**. Artigo para jornal eletrônico na Internet, NE10 – Sistema Jornal do Commercio de Comunicação, Recife, Brasil (2013e).

Alencar, M. S., **A Anatel faz o Jogo das Operadoras de Telecomunicações**. Artigo para jornal eletrônico na Internet, NE10 – Sistema Jornal do Commercio de Comunicação, Recife, Brasil (2014a).

Alencar, M. S., **A Corrupção é Endêmica no Setor Privado**. Artigo para jornal eletrônico na Internet, NE10 – Sistema Jornal do Commercio de Comunicação, Recife, Brasil (2014b).

Alencar, M. S., **As empresas Virtuais Exploram o Trabalho Real dos Usuários**. Artigo para jornal eletrônico na Internet, NE10 – Sistema Jornal do Commercio de Comunicação, Recife, Brasil (2014c).

Alencar, M. S., **Como Aprender com Filósofos e Cientistas**. Artigo para jornal eletrônico na Internet, NE10 – Sistema Jornal do Commercio de Comunicação, Recife, Brasil (2014d).

Alencar, M. S., **Como o Dinheiro das Empresas Privatizadas Sai do País**. Artigo para jornal eletrônico na Internet, NE10 – Sistema Jornal do Commercio de Comunicação, Recife, Brasil (2014e).

Alencar, M. S., **Empresas Virtuais Violam Regras da Economia Clássica**. Artigo para jornal eletrônico na Internet, NE10 – Sistema Jornal do Commercio de Comunicação, Recife, Brasil (2014f).

Alencar, M. S., **No Final Sera o Caos**. Artigo para jornal eletrônico na Internet, NE10 – Sistema Jornal do Commercio de Comunicação, Recife, Brasil (2014g).

Alencar, M. S., *Teoria de Conjuntos, Medida e Probabilidade*. Editora Érica Ltda., ISBN 978-85-365-0715-6, São Paulo, Brasil (2014h).

Alencar, M. S., **O Atual Sistema Feudal**. Artigo para jornal eletrônico na Internet, NE10 – Sistema Jornal do Commercio de Comunicação, Recife, Brasil (2015a).

Alencar, M. S., **O Fim do Sistema Feudal e o Início da Democracia**. Artigo para jornal eletrônico na Internet, NE10 – Sistema Jornal do Commercio de Comunicação, Recife, Brasil (2015b).

Alencar, M. S., **The Strange Economic Behavior of the Information Era**. *Revista de Tecnologia da Informação e Comunicação*, 5(1):39–43 (2015c).

Alencar, M. S., **Truques para Fazer Elisão Fiscal**. Artigo para jornal eletrônico na Internet, NE10 – Sistema Jornal do Commercio de Comunicação, Recife, Brasil (2015d).

Alencar, M. S., **A Economia Feudal na Era da Tecnologia da Informação e Comunicação**. *Revista Ciência & Trópico – Fundação Joaquim Nabuco (ISSN 0304-2685)*, 40(1):Publicação eletrônica (2016a).

Alencar, M. S., **Sobre Impostos e Tarifas**. Artigo para jornal eletrônico na Internet, NE10 – Sistema Jornal do Commercio de Comunicação, Recife, Brasil (2016b).

Alencar, M. S., **Distribuição de Renda na Economia**. Artigo para jornal eletrônico na Internet, NE10 – Sistema Jornal do Commercio de Comunicação, Recife, Brasil (2017a).

Alencar, M. S., **Sobre Tulipas e Moedas Virtuais**. Artigo para jornal eletrônico na Internet, NE10 – Sistema Jornal do Commercio de Comunicação, Recife, Brasil (2017b).

Alencar, M. S., **Todos Somos Empregados dos Bancos**. Artigo para jornal eletrônico na Internet, NE10 – Sistema Jornal do Commercio de Comunicação, Recife, Brasil (2017c).

Alencar, M. S., **A Riqueza Singular: A Semelhança entre a Economia e o Universo**. Artigo para jornal eletrônico na Internet, NE10 – Sistema Jornal do Commercio de Comunicação, Recife, Brasil (2018a).

Alencar, M. S., **O Material e o Virtual**. Artigo para jornal eletrônico na Internet, NE10 – Sistema Jornal do Commercio de Comunicação, Recife, Brasil (2018b).

Alencar, M. S., **Sobre Quadros e Computadores**. Artigo para jornal eletrônico na Internet, NE10 – Sistema Jornal do Commercio de Comunicação, Recife, Brasil (2018c).

Alencar, M. S., **Uma Análise dos Resultados Econômicos e Fiscais da Privatização das Empresas Estatais de Telecomunicações**. *Revista Ciência & Trópico – Fundação Joaquim Nabuco (ISSN 2526-9372)*, 42(1):187–202 (2018d).

Alencar, M. S. and Alencar, R. T., *Probability Theory*. Momentum Press, LLC, ISBN-13: 978-1-60650-747-6 (print), New York, USA (2016).

Alencar, M. S. and da Rocha Jr., V. C., *Communication Systems*. Springer, ISBN 0-387-25481-1, Boston, USA (2005).

Alvaredo, F., Chancel, L., Piketty, T., Saez, E., and Zucman, G., *World Inequality Report*. World Inequality Lab, Paris School of Economics, Paris, France (2018).

Alves, I. M., **Glossário de Termos Neológicos da Economia**. Technical report, São Paulo, Brasil (2001).

Balckburn, S., *Dicionário Oxford de Filosofia*. Jorge Zahar Editor, Rio de Janeiro, Brasil (1997).

Bayes, T., **An essay towards solving a problem in the doctrine of chance**. *Philosophical Transactions of the Royal Society of London*, (53):370–418 (1763).

Bellone, B. and Gautier, E., **Predicting economic downturns through a financial qualitative hidden Markov model**. Technical report (2004).

Berthoud, A., **A História do Pensamento Econômico e sua herança filosófica**. *Econômica*, 3(1):63–67 (2000).

Blomqvist, A., Fried, J., Wonnacott, P., and Wonnacott, R., *An Introduction to Microeconomics*. MacGraw-Hill Ryerson Limited, Toronto, Canada (1994).

Blomqvist, A., Wonnacott, P., and Wonnacott, R., *An Introduction to Microeconomics*. MacGraw-Hill Ryerson Limited, Toronto, Canada (1987).

Bornstein, M. and Fusfeld, D. R., *The Soviet Economy – A Book of Readings*. Richard D. Irwin, Inc., Illinois, United States (1970).

Braga, F., Lopes, W., and Alencar, M. S., **Cognitive Vehicular Networks: An Overview**. *Journal of Procedia Computer Science (ISSN 1877-0509)*, 65(1):Electronic publication (2015).

Bresser-Pereira, L. C., **Modelos de Estado desenvolvimentista**. *Revista de Economia*, 40(73):231–156 (2019).

Bronstein, I. and Semendiaev, K., *Manual de Matemática*. Editora Mir, Moscou (1979).

Brzezniak, Z. and Zastawniak, T., *Basic Stochastic Processes*. Springer, London, Great Britain (2006).

BTG Pactual Digital, **IGP-M: o que é o índice, tabela anual, mensal e acumulado**. Internet site, https://www.btgpactualdigital.com/blog/fin ancas/igp-m-o-que-e-o-indice (2020).

Bureau, C., **Statistics of U.S. Businesses (SUSB) Main**. Internet webpage, Available at https://www.census.gov/econ/susb/ (2011).

Camacho, T. S. and da Silva, G. J. C., **Criptoativos: Uma Análise do Comportamento e da Formação do Preço do Bitcoin**. *Revista de Economia*, 39(68):1–26 (2018).

Capiński, M. and Kopp, P. E., *Measure, Integral and Probability*. Springer-Verlag London Ltd., New York, USA (2005).

Castro, E. R. S., Alencar, M. S., and da Fonseca, I. E., **Probability Density Functions of the Packet Length for Computer Networks with Bimodal Traffic**. *International Journal of Computer Networks & Communications (IJCNC)*, 5(3) (2013).

Chacholiades, M., *Microeconomics*. MacMillan Publishing Company, New York, United States (1986).

Chowdhury, S. R., **Wealth inequality, entrepreneurship and industrialization**. *Journal of Economics, Springer-Verlag*, 108:81–102 (2013).

Clemens, C. and Heinemann, M., **On the Effects of Redistribution on Growth and Entrepreneurial Risk-taking**. *Journal of Economics, Springer-Verlag*, 88(2):131–158 (2006).

Comin, A., **Glossário de Termos de Economia Industrial**. Internet site, https://www.pucsp.br/~acomin/economes/glosglob.html (2020).

Cournot, A. A., *Researches into the Mathematical Principles of the Theory of Wealth*. Chez L. Hachette, Paris, France (1838).

Cournot, A. A., *Researches into the Mathematical Principles of the Theory of Wealth, Translation by Nathaniel T. Bacon*. The MacMillan Company, London, United Kingdom (1897).

Currais, L., **From the Malthusian regime to the demographic transition: Contemporary research and beyond**. *Econômica*, 3(1):75–101 (2000).

da Cruz e Mello Moraes, M. V., *Poemas, Sonetos e Baladas*. Edições Gavetas, São Paulo (1946).

Darwin, C., *On the Origin of Species*. Penguin Books, New York, USA (2009).

de Azevedo, L. F. and cei ção, O. A. C. C., **Incerteza e não-ergodicidade: as perspectivas das vertentes institucionalistas**. *Revista de Economia*, 38(67):1–24 (2017).

de Carvalho, F. G. P., *Do Processo Inflacionário, como Subsídio Estatal à Concentração da Renda Privada*. Edição do Autor, Recife, Brasil (2012).

de Souza, J., *A Elite do Atraso – da Escravidão à Lava Jato*. Casa da Palavra/Leya, Rio de Janeiro, Brasil (2017).

de Souza Silva, G., **As controvérsias no marxismo sobre a compreensão dos atributos e funções do dinheiro**. *Revista de Economia*, 39(69):1–26 (2018).

Dreifuss, R. A., *1964, A Conquista do Estado: Ação Política, Poder e Golpe de Classe*. Editora Vozes, Petrópolis, Brasil (1981).

Dória, P., *O Príncipe da Privataria*. Geração Editorial, São Paulo, Brasil (2011).

Dunham, W., *Journey through Genius – The Great Theorems of Mathematics*. Penguin Books, New York, USA (1990).

Economia.net, **Dicionário de Economia**. Internet site, https://www.econom iabr.net/dicionario/economes_uvz.html (2020).

Editora Melhoramentos Ltda., **Dicionário Brasileiro da Língua Portuguesa**. Internet site, http://michaelis.uol.com.br/busca?id=GVKd (2020).

Einstein, A., **Investigations on the Theory of the Brownian Movement**. *Annalen der Phys*, 17(549) (1905).

Estadão, **E&N Glossário**. Internet site, https://economia.estadao.com.br/gl ossario/ (2020).

Farias, T. A., **Teoria da estabilidade do equilíbrio competitivo em mercado múltiplo numa economia Arrow & Debreu**. *Revista de Economia*, 36(1):67–100 (2010).

Fischer, I., **Cournot and Mathematical Economics**. *The Quaterly Journal of Economics*, 12(2):119–138 (1898).

Folha de São Paulo, **Dicionário de Economês**. Internet site, https://m.folha. uol.com.br/mercado/2013/01/1218059-confira-o-dicionario-de-economes -e-tire-duvidas-sobre-os-termos.shtml (2020).

Frazão, A., Dias, E., Gonçalves, J., Pereira, L., and Marques, R., **Dicionário Financeiro**. Internet site, https://www.dicionariofinanceiro.com/ (2020).

Furtado, C., *Formação Econômica do Brasil*. Companhia Editora Nacional, São Paulo, Brasil (1998).

Gal-Or, B., *Cosmology, Physics, and Philosophy*. Springer-Verlag New York Inc., New York, USA (1981).

Galbraith, J. K., *The Affluent Society*. New American Library, Peguin Books, Markham, Canada (1984).

Gleeson, J., *O Inventor do Papel*. Editora Rocco Ltda., Rio de Janeiro, Brasil (1999).

Greenlaw, S. A. and Shapiro, D., **Principles of macroeconomics 2e**. [Online; accessed 01-June-2020] (2018).

Hashemi, H., **Principles of Digital Indoor Radio Propagation**. In *IASTED International Symposium on Computers, Electronics, Communication and Control*, pp. 271–273, Calgary, Canada (1991).

Heilbroner, R. L., *The Wordly Philosophers*. Touchtone, Simon and Schuster, New York, USA (1986).

Heilbroner, R. L. and Thurow, L. C., *Economics Explained*. Prentice-Hall, Inc., Englewood Cliffs, New Jersey, USA (1982).

Hodgson, G., **A evolução das Instituições: Uma agenda para pesquisa teórica futura**. *Econômica*, 3(1):97–135 (2001).

InfoMoney, **GLossário**. Internet site, https://www.infomoney.com.br/glossario (2020).

Itô, K., **"Stochastic integral"**. *Proceedings of the Imperial Academy*, 8(20):519–524 (1944).

Itô, K., **"On a Stochastic Integral Equation"**. *Proceedings of the Imperial Academy*, 2(22):32–35 (1946).

Jairo Laser Proclanoy and Paulo Cesar Delaytl Motta, **Inovação tecnológica: a privatização após o uso das "moedas podres"**. *Revista de Administração de Empresas*, 32(5):48–60 (1992).

Jaqueline Guimarães Santos, **"Production Planning and Control of Hawaiian: A Case Study in Alpargatas of Campina Grande"**. *Revista Gestão Industrial*, 9(3):623–640 (2013).

Joyce, J., **Bayes' theorem**. In Zalta, E. N., editor, *The Stanford Encyclopedia of Philosophy*. Fall 2008 edition (2008).

Jr., A. R., *A Privataria Tucana*. Geração Editorial, São Paulo, Brasil (2011).

Kennedy, R. S., *Fading Dispersive Communication Channels*. Wiley-Interscience, New York (1969).

Koutsoyannis, A., *Theory of Econometrics*. MacMillan Education Ltd., London, United Kingdom (1977).

Krane, K. S., *Modern Physics*. John Wiley & Sons, New York, United States (1999).

Kreinin, M. E., *International Economics – A Policy Approach*. Harcourt Brace Jovanovich, Publishers, Orlando, USA (1987).

Kurtz, M., *Engineering Economics for Professional Engineers Examinations*. McGraw-Hill Book Company, New York (1975).

Lecours, M., Chouinard, J.-Y., Delisle, G. Y., and Roy, J., **Statistical Modeling of the Received Signal Envelope in a Mobile Radio Channel**. *IEEE Transactions on Vehicular Technology*, 37(4):204–212 (1988).

Little, J. D. C., **Little's Law as Viewed on Its 50th Anniversary**. *Operations Research*, 59(3):536–549 (2011).

Magalhães, M. N., *Probabilidade e Variáveis Aleatórias*. Editora da Universidade de São Paulo, São Paulo, Brasil (2006).

Mankiw, N. G., *Microeconomics*. Worth Publishers, New York, United States (2010).

Marques, M., *Teoria da Medida*. Editora da Unicamp, Campinas, Brasil (2009).

Marx, K., *Capital: Volume I – A Critique of Political Economy*. Penguin Books, New York, USA (2004).

Mattei, L. and dos Santos Júnior, J. A., **Industrialização e Substituição de Importações no Brasil e na Argentina: Uma Análise Histórica Comparada**. *Revista de Economia*, 35(1):93–115 (2009).

Medeiros, F. J. M. and do Val Cosentino, D., **Celso Furtado e Raúl Prebisch frente à crise do desenvolvimentismo da década de 1960**. *Revista de Economia*, 41(74):150–179 (2020).

Mirow, K. R., *A Ditadura dos Carteis*. Editora Civilização Brasileira, Rio de Janeiro (1978).

Mises, R. V., **Über die Ganzzahligkeit der Atomgewicht and Verwandte Fragen**. *Physikal. Z.*, 19:490–500 (1918).

Morier, B. and Teles, V. K., **A Time-Varying Markov-Switching Model for Economic Growth**. Working paper 018 (2011).

Movellan, J. R., **"Tutorial on Stochastic Differential Equations"**. Technical report, MPLab Tutorials Version 06.1 (2011).

Neal H. Shepherd, Editor, **Received Signal Fading Distribution**. *IEEE Transactions on Vehicular Technology*, 37(1):57–60 (1988).

Niall Ksihtainy, E. C., *O Livro da Economia*. Editora Globo, São Paulo, Brasil (2013).

Norris, J. R., *Makov Chains*. Cambridge University Press, New York, USA (1997).

Oksendal, B., *Stochastic Differential Equations — An Introduction with Applications*. Springer-Verlag, Berlin, Germany (2013).

Papoulis, A., **Random Modularion: a Review**. *IEEE Transactions on Accoustics, Speech and Signal Processing*, 31(1):96–105 (1983).

Pareto, V., *Cours D'ècomomie Politique – Professé a L'Université Lausanne, Tome Premier*. F. Rouge, Éditeur, Paris, France (1896).

Pareto, V., *Cours D'ècomomie Politique – Professé a L'Université Lausanne, Tome Second*. F. Rouge, Éditeur, Paris, France (1897).

Pareto, V., *Cours d'Économie Politique: Nouvelle édition par G.-H. Bousquet et G. Busino*. Librairie Droz, Geneve, Switzerland (1964).

Parker, D., **The uk's privatisation experiment: The passage of time permits a sober assessment**. Working paper, https://www.SSRN.com (2003).

Piketty, T., *Capital in the Twenty-First Century*. Harvard University Press, Cambridge, USA (2014).

Pivetta, M. and de Oliveira Andrade, R., **Quando a luz se curvou**. Sítio da internet, https://revistapesquisa.fapesp.br/quando-a-luz-se-curvou/ (2019).

Plato, *The Republic of Plato*. Oxford University Press, London, United Kingdom (1970).

Proakis, J. G., *Digital Communications*. McGraw-Hill Book Company, New York (1990).

Póvoa, L. M. C. and Monsueto, S. E., **Tamanho das Empresas, Interação com Universidades e Inovação**. *Revista de Economia*, 37(Especial):09–24 (2011).

Rappaport, T. S., **Indoor Radio Communications for Factories of the Future**. *IEEE Communications Magazine*, pp. 15–24 (1989).

Reiβ, M., **"Stochastic Differential Equations"**. Lecture notes, Institute of Applied Mathematics, University of Heidelberg (2007).

Rényi, A., *Probability Theory*. Dover Publications, Inc., New York, USA (2007).

Ricardo, D., *Princípios de Economia Política e Tributação*. Editora Nova Cultural, São Paulo, Brasil (1996).

Rocha, F., **Composição do crescimento dos serviços na economia brasileira: uma análise da matriz insumo-produto (1985-1992)**. *Econômica*, 1(2):107–130 (1999).

Rodrigues, R. V. and Teixeira, E. C., **Gasto Público e Crescimento Econômico no Brasil: Uma Análise Comparativa dos Gastos das Esferas de Governo**. *Revista Brasileira de Economia*, 64(4):423–438 (2010).

Roncayolo, M., *Histoire du Monde Contemporain*. Bordas, Paris (1973).

Rosenthal, J. S., *A First Look at Rigorous Probability Theory*. World Scientific Publishing Co. Pte. Ltd., Singapore (2000).

Roser, M., **Global economic inequality**. *Our World in Data*. https://ourworldindata.org/global-economic-inequality (2013).

Rousseau, J.-J., *The Social Contract and Discourses*. David Campbell Publishers Ltd., London, Great Britain (1973).

Rumjanek, F. D., *AB Initio – Origem da Vida e Evolução*. Vieira & Lent, Rio de Janeiro, Brasil (2009).

Sabine, G. H. and Thorson, T. L., *A History of Political Theory*. Hinsdale, United States (1973).

SBA, **The U.S. Small Businesses Administration (SBA)**. Internet webpage, United States Government, Available at `https://http://www.sba.gov/` (2014).

Schumacher, E. F., *Small is Beautiful – A Study of Economics as if People Mattered*. Abacus, Sphere Books Ltd., London, Great Britain (1990).

Schwartz, M., *Information Transmission, Modulation, and Noise*. McGraw-Hill, New York (1970).

Schwartz, M., Bennett, W., and Stein, S., *Communication Systems and Techniques*. McGraw-Hill, New York (1966).

Siegel, J. G. and Shim, J. K., *Dictionary of Accounting Terms*. Barron's Educationla Series, Inc., New York, United States (1987).

Smith, A., *The Wealth of Nations – Books I-III*. Penguin Books Ltd., London, Engaland (1986).

Soares, M. M., Barbosa, R. F., de Sousa Oliveira, W. A., de Macedo, P. P., and Duarte, L. L. Q., **"Análise do Sistema de Produção em uma Empresa Calçadista: Caso da Alpargatas S.A."**. In *XXXI Encontro Nacional de Engenharia de Produção*, pp. 1–14, Belo Horizonte, Brasil (2011).

Tannenbaum, E. R., *A History of World Civilizations*. John Wiley & Sons, Inc., New York, USA (1973).

Utsch, M., **"Relatório da Administração 2013"**. Relatório técnico, Alpargatas S/A (2014).

Utsch, M. and Letiere, J. R., **"Divulgação dos Resultados 4T13 e 2013"**. Relatório técnico, Alpargatas S/A (2014).

Vaz, L., *Sanguessugas do Brasil*. Geração Editorial, São Paulo, Brasil (2012).

Veljanovski, C. G., **Privatization in britain – the institutional and constitutional issues**. Article, http://scholarship.law.marquette.edu/mulr (1988).

Ventura, L., **Wealth Distribution and Income Inequality by Country 2018**. *Global Finance Magazine* (2018).

Vinod Thomas, E., *World Development Report*. Oxford University Press, London, Great Britain (1991).

Weber, J. E., *Mathematical Analysis – Business and Economic Applications*. Harper & Row, Publishers, New York, USA (1982).

Weibull, W., **A statistical distribution function of wide applicability**. *Journal of Applied Mechanics*, 18:293–296 (1951).

Wikipédia, **Capitalismo – Wikipédia, a enciclopédia livre**. [Online; accessed 24-May-2020] (2020a).

Wikipédia, **Richard Cantillon — wikipédia, a enciclopédia livre**. [Online; accessed 08-fevereiro-2020] (2020b).

Wikipedia contributors, **Pareto distribution — Wikipedia, the free encyclopedia**. [Online; accessed 17-June-2021] (2021).

World Economic Forum, *The Inclusive Development Index 2018 – Summary and Data Highlights*. World Economic Forum, Geneva, Switzerland (2018).

Index

About the Author

Marcelo Sampaio de Alencar was born in Serrita, Brazil, in 1957. He received his bachelor degree in electrical engineering, from the Federal University of Pernambuco (UFPE), Brazil, 1980, his master degree in electrical engineering, from the Federal University of Paraiba (UFPB), Brazil, 1988 and his Ph.D. from the University of Waterloo, Department of Electrical and Computer Engineering, Canada, 1993. He has 40 years of engineering experience, and 30 years as an IEEE Member, currently as a senior member. Between 1982 and 1984, he worked for the State University of Santa Catarina (UDESC). From 1984 to 2003, he worked for the Department of Electrical Engineering, Federal University of Paraiba, where he was a Full Professor and supervised more than 60 graduate and several undergraduate students. From 2003 to 2017, he was Chair Professor at the Department of Electrical Engineering, Federal University of Campina Grande, Brazil. He also spent some time working for MCI-Embratel and the University of Toronto, as Visiting Professor. He was Visiting Chair Professor at the Department of Electrical Engineering, Federal University of Bahia. He is now Visiting Researcher at the Senai Cimatec, Salvador, Bahia, Brasil.

He is the founder and president of the Institute for Advanced Studies in Communications (Iecom). He has been awarded several scholarships and grants, including three scholarships and several research grants from the Brazilian National Council for Scientific and Technological Research (CNPq), two grants from the IEEE Foundation, a scholarship from the University of Waterloo, a scholarship from the Federal University of Paraiba, an achievement award for contributions to the Brazilian Telecommunications Society (SBrT), an academic award from the Medicine College of the Federal University of Campina Grande (UFCG), and an achievement award from the College of Engineering of the Federal University of Pernambuco, during its 110th year celebration. He is a laureate of the 2014 Attilio Giarola Medal.

He published over 450 engineering and scientific papers and 26 books: *Music Science, Linear Electronics*, *Modulation Theory, Scientific Style in English*, and *Cellular Network Planning*, by River Publishers, *Spectrum Sensing Techniques and Applications, Information Theory, and Probability Theory*, by Momentum Press, *Information, Coding and Network Security* (in Portuguese), by Elsevier, *Digital Television Systems*, by Cambridge, *Communication Systems*, by Springer, *Principles of Communications* (in Portuguese), by Editora Universitária da UFPB, *Set Theory, Measure and Probability, Computer Networks Engineering, Electromagnetic Waves and Antenna Theory, Probability and Stochastic Processes, Digital Cellular Telephony, Digital Telephony, Digital Television and Communication Systems* (in Portuguese), by Editora Érica Ltda, *History of Communications in Brazil, History, Technology and Legislation of Communications, Connected Sex, Scientific Diffusion, Soul Hicups* (in Portuguese), by Epgraf Gráfica e Editora. He also wrote several chapters for 11 books. His biography is included in the following publications: *Who's Who in the World* and *Who's Who in Science and Engineering*, by Marquis Who's Who, New Providence, USA.

Marcelo S. Alencar has contributed in different capacities to the following scientific journals: Editor of the *Journal of the Brazilian Telecommunication Society*; Member of the International Editorial Board of the *Journal of Communications Software and Systems* (JCOMSS), published by the Croatian Communication and Information Society (CCIS); Member of the Editorial Board of the *Journal of Networks* (JNW), published by Academy Publisher; founder and editor-in-chief of the *Journal of Communication and Information*

Systems (JCIS), special joint edition of the IEEE Communications Society (ComSoc) and SBrT. He is a member of the SBrT-Brasport Editorial Board. He has been involved as a volunteer with several IEEE and SBrT activities, including being a member of the Advisory or Technical Program Committee in several events. He served as a member of the IEEE Communications Society Sister Society Board and as liaison to Latin America Societies. He also served on the Board of Directors of IEEE's Sister Society SBrT. He is a Registered Professional Engineer. He was a columnist of the traditional Brazilian newspaper Jornal do Commercio for two decades, and he was vice-president external relations of SBrT. He is a member of the IEEE, IEICE, in Japan, SBrT, SBMO, SBPC, ABJC and SBEB, in Brazil. He studied acoustic guitar at the Federal University of Paraiba, and keyboard and bass at the music school Musidom. He is composer and percussionist of the carnival club *Bola de Ferro*, in Recife, Brazil.